Understanding Antimicrobial Drug Resistance

Understanding Antimicrobial Drug Resistance

Scott Gibbons

STATES
ACADEMIC PRESS
www.statesacademicpress.com

Published by States Academic Press,
109 South 5th Street,
Brooklyn, NY 11249, USA
www.statesacademicpress.com

Understanding Antimicrobial Drug Resistance
Scott Gibbons

International Standard Book Number: 978-1-63989-535-9 (Hardback)

Cataloging-in-Publication Data

Understanding antimicrobial drug resistance / Scott Gibbons.
p. cm.
Includes bibliographical references and index.
ISBN 978-1-63989-535-9
1. Drug resistance in microorganisms. 2. Anti-infective agents. 3. Microorganisms--Effect of drugs on.
4. Drug resistance. I. Gibbons, Scott.

QR177 .U53 2022
616.904 1--dc23

Table of Contents

Preface

This book is a culmination of my many years of practice in this field. I attribute the success of this book to my support group. I would like to thank my parents who have showered me with unconditional love and support and my peers and professors for their constant guidance.

Antimicrobial drug resistance refers to the ability of microbes such as bacteria, fungi virus and parasite, to resist the effects of drugs and medications, which were used to treat them earlier. The type of antimicrobial resistance, in which microbes resist to multiple antimicrobials, is known as multiple drug resistance. There are multiple causes for such resistance, such as natural (biological) causes, inappropriate use of antibiotics, water pollution, inadequate diagnostics, etc. All microbes have the tendency to develop resistance. Fungal resistance, antibiotic resistance, antiviral resistance, antiprotozoal resistance, etc. are some forms of antimicrobial resistance. This book is a valuable compilation of topics, ranging from the basic to the most complex theories and principles of antimicrobial drug resistance. It explores all the important aspects of the subject in the present day scenario. Coherent flow of topics, student-friendly language and extensive use of examples make this book an invaluable source of knowledge.

The details of chapters are provided below for a progressive learning:

Chapter - Introduction

The ability of a microbe to resist the effects of the medication is known as antimicrobial drug resistance. Its major causes include selective pressure, mutation and gene transfer. This is an introductory chapter which will briefly introduce all the significant aspects of antimicrobial drug resistance.

Chapter - Types of Antimicrobial Drug

Antimicrobial drugs are used for treating a microbial infection. There are various types of antimicrobial drugs which can be named as antiviral drug, antifungal medication, antiparasitics and antibiotic medication. The topics elaborated in this chapter will help in gaining a better perspective about these types of antimicrobial drugs.

Chapter - Antimicrobial Therapies

Antimicrobial therapy helps in killing and inhibiting the growth of bacteria, fungi and protozoans. Empiric antimicrobial therapy and antimicrobial chemotherapy are the most important antimicrobial therapies. This chapter closely examines the key concepts related to antimicrobial therapy to provide an extensive understanding of the subject.

Chapter - Multiple Drug Resistance

The antimicrobial resistance which is shown by a species of microorganisms to multiple drugs is referred to as multiple drug resistance. Multidrug tolerance and resistance in bacteria and fungi are studied within it. The topics elaborated in this chapter will help in gaining a better perspective about the subject of multiple drug resistance.

Chapter - Diverse Aspects of Antimicrobial Drug Resistance

There are many important concepts that are studied under the domain of antimicrobial drug resistance such as horizontal gene transfer, eagle effect, selectivity factor and r-factor. This chapter has been carefully written to provide an easy understanding of these related concepts of antimicrobial drug resistance.

Scott Gibbons

1

Introduction

The ability of a microbe to resist the effects of the medication is known as antimicrobial drug resistance. Its major causes include selective pressure, mutation and gene transfer. This is an introductory chapter which will briefly introduce all the significant aspects of antimicrobial drug resistance.

ANTIMICROBIAL DRUG

An antimicrobial is a substance that kills or inhibits the growth of microorganisms such as bacteria, fungi, or protozoans.

An antimicrobial is a substance that kills or inhibits the growth of microorganisms such as bacteria, fungi, or protozoans. Antimicrobial drugs either kill microbes (microbiocidal) or prevent the growth of microbes (microbiostatic). Disinfectants are antimicrobial substances used on non-living objects or outside the body.

The history of antimicrobials begins with the observations of Pasteur and Joubert, who discovered that one type of bacteria could prevent the growth of another. They did not know at that time that the reason one bacterium failed to grow was that the other bacterium was producing an antibiotic. Technically, antibiotics are only those substances that are produced by one microorganism that kill, or prevent the growth, of another microorganism. The term antibiotic is used to refer to almost any drug that attempts to rid your body of a bacterial infection. Antimicrobials include not just antibiotics, but synthetically formed compounds as well.

The discovery of antimicrobials like penicillin and tetracycline paved the way for better health for millions around the world. Before penicillin became a viable medical treatment in the early 1940s, no true cure for gonorrhea, strep throat, or pneumonia existed. Patients with infected wounds often had to have a wounded limb removed, or face death from infection. Now, most of these infections can be cured easily with a short course of antimicrobials.

However, with the development of antimicrobials, microorganisms have adapted and become resistant to previous antimicrobial agents. The old antimicrobial technology was based either on poisons or heavy metals, which may not have killed the microbe completely, allowing the microbe to survive, change, and become resistant to the poisons and/or heavy metals.

Antimicrobial nanotechnology is a recent addition to the fight against disease-causing organisms, replacing heavy metals and toxins, and may some day be used as a viable alternative.

Infections that are acquired during a hospital visit are called "hospital acquired infections" or nosocomial infections. Similarly, when the infectious disease is picked up in the non-hospital setting, it is considered "community acquired".

Synthetic agents include: sulphonamides, cotrimoxazole, quinolones, anti-virals, anti-fungals, anti-cancer drugs, anti-malarials, anti-tuberculosis drugs, anti-leprotics, and anti-protozoals.

Sulfonamide or sulphonamide is the basis of several groups of drugs. The original antibacterial sulfonamides (sometimes called sulfa drugs or sulpha drugs) are synthetic antimicrobial agents that contain the sulfonamide group. Some sulfonamides are also devoid of antibacterial activity, e.g., the anticonvulsant sultiame. The sulfonylureas and thiazide diuretics are newer drug groups based on the antibacterial sulfonamides.

Sulfa allergies are common, and medications containing sulfonamides are prescribed carefully. It is important to make a distinction between sulfa drugs and other sulfur-containing drugs and additives, such as sulfates and sulfites, which are chemically unrelated to the sulfonamide group and do not cause the same hypersensitivity reactions seen in the sulfonamides.

In bacteria, antibacterial sulfonamides act as competitive inhibitors of the enzyme dihydropteroate synthetase (DHPS), an enzyme involved in folate synthesis. As such, the microorganism will be "starved" of folate and die.

The sulfonamide chemical moiety is also present in other medications that are not antimicrobials, including thiazide diuretics (including hydrochlorothiazide, metolazone, and indapamide, among others), loop diuretics (including furosemide, bumetanide, and torsemide), sulfonylureas (including glipizide, glyburide, among others), and some COX-2 inhibitors (e.g., celecoxib), and acetazolamide.

Naturally Occurring Antimicrobial Drugs: Antibiotics

An antimicrobial is a substance that kills or inhibits the growth of microorganisms such as bacteria, fungi, or protozoans.

An antimicrobial is a substance that kills or inhibits the growth of microorganisms bacteria, fungi, or protozoans. Antimicrobial drugs either kill microbes (microbiocidal) or prevent the growth of microbes (microbiostatic). Disinfectants are antimicrobial substances used on non-living objects or outside the body.

The discovery of antimicrobials, like penicillin and tetracycline, paved the way for better health for millions of people around the world. Before penicillin became a viable medical treatment in the early 1940's, no true cure for gonorrhea, strep throat, or pneumonia existed. Patients with infected wounds often had to have a wounded limb removed or face death from infection. Now, most of these infections can be cured easily with a short course of antimicrobials.

However, with the development of antimicrobials, microorganisms have adapted and become resistant to previous antimicrobial agents. The old antimicrobial technology was based either on poisons or heavy metals, which may not have killed the microbe completely. This allowed the microbe to survive, change, and become resistant to the poisons and/or heavy metals.

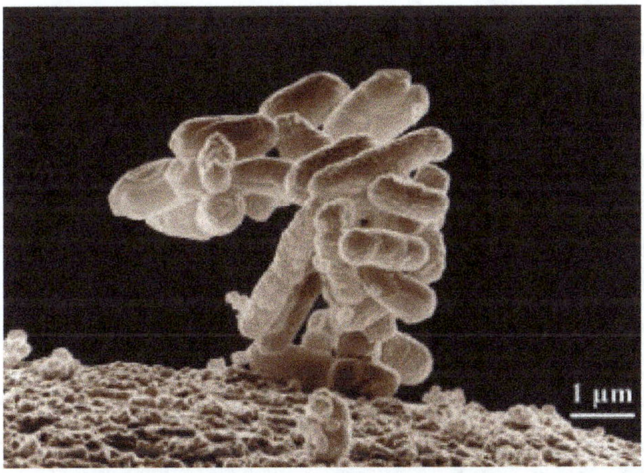

A cluster of Escherichia coli Bacteria magnified 10,000 times.

Antimicrobial nanotechnology is a recent addition to the fight against disease causing organisms. It replaces heavy metals and toxins and may someday be a viable alternative.

Infections that are acquired during a hospital visit are called "hospital acquired infections" or nosocomial infections. Similarly, when the infectious disease is picked up in the non-hospital setting it is considered "community acquired."

There are mainly two classes of antimicrobial drugs: those obtained from natural sources (i.e. beta-lactam) antibiotic (such as penicillins, cephalosporins) or protein synthesis inhibitors (such as aminoglycosides, macrolides, tetracyclines, chloramphenicol, polypeptides); and synthetic agents.

A β-lactam (beta-lactam) ring is a four-membered lactam. A lactam is a cyclic amide. It is named as such, because the nitrogen atom is attached to the β-carbon relative to the carbonyl. The simplest β-lactam possible is 2-azetidinone.

A protein synthesis inhibitor is a substance that stops or slows the growth or proliferation of cells by disrupting the processes that lead directly to the generation of new proteins. While a broad interpretation of this definition could be used to describe nearly any antibiotic, in practice, it usually refers to substances that act at the ribosome level (either the ribosome itself or the translation factor), taking advantage of the major differences between prokaryotic and eukaryotic ribosome structures. Toxins such as ricin also function via protein synthesis inhibition. Ricin acts at the eukaryotic 60S.

In general, protein synthesis inhibitors work at different stages of prokaryotic mRNA translation into proteins, like initiation, elongation (including aminoacyl tRNA entry, proofreading, peptidyl transfer, and ribosomal translocation), and termination. Rifamycin inhibits prokaryotic DNA transcription into mRNA by inhibiting DNA-dependent RNA polymerase by binding its beta-subunit. Linezolid acts at the initiation stage probably by preventing the formation of the initiation complex, although the mechanism is not fully understood.

Tetracyclines and Tigecycline (a glycylcycline related to tetracyclines) block the A site on the ribosome, preventing the binding of aminoacyl tRNAs. Aminoglycosides, among other potential mechanisms of action, interfere with the proofreading process, causing increased rate of error in synthesis with premature termination. Chloramphenicol blocks the peptidyl transfer step of elongation on the 50S ribosomal subunit in both bacteria and mitochondria. Macrolides (as well as inhibiting ribosomal translocation and other potential mechanisms) bind to the 50s ribosomal subunits, inhibiting peptidyl transfer. Quinupristin/dalfopristin act synergistically, with dalfopristin, enhancing the binding of quinupristin as well as inhibiting peptidyl transfer. Quinupristin binds to a nearby site on the 50S ribosomal subunit and prevents elongation of the polypeptide. It also causes incomplete chains to be released. Macrolides, clindamycin, and aminoglycosides (with all these three having other potential mechanisms of action as well) have evidence of inhibition of ribosomal translocation. Fusidic acid prevents the turnover of elongation factor G (EF-G) from the ribosome. Macrolides and clindamycin (both also having other potential mechanisms) cause premature dissociation of the peptidyl-tRNA from the ribosome. Puromycin has a structure similar to that of the tyrosinyl aminoacyl-tRNA. Therefore, it binds to the ribosomal A site and participates in peptide bond formation, producing peptidyl-puromycin. However, it does not engage in translocation and quickly dissociates from the ribosome, causing a premature termination of polypeptide synthesis.

Beta-lactam Antibiotics: Penicillins and Cephalosporins

The β-lactam ring is part of the core structure of several antibiotic families. A β-lactam (beta-lactam) ring, is a four-membered lactam. It is named as such, because the nitrogen atom is attached to the β-carbon relative to the carbonyl. The simplest β-lactam possible is 2-azetidinone.

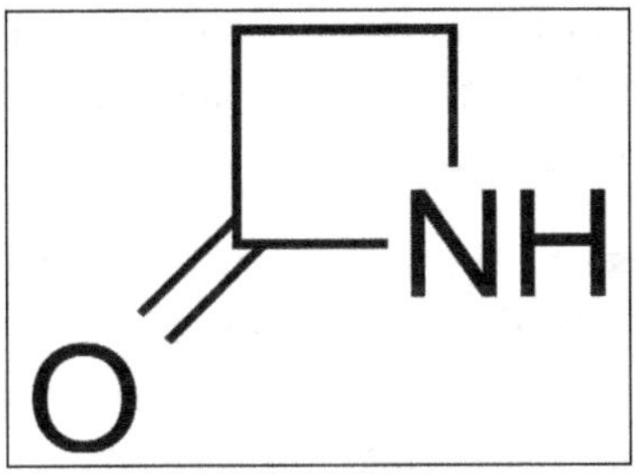

β-Lactam: β-Lactam ring is a four-membered lactam.

The β-lactam ring is part of the core structure of several antibiotic families, the principal ones being the penicillins, cephalosporins, carbapenems, and monobactams, which are, therefore, also called β-lactam antibiotics. Nearly all of these antibiotics work by inhibiting bacterial cell wall biosynthesis. This has a lethal effect on bacteria. Bacteria do, however, contain within their populations, in smaller quantities, bacteria that are resistant against β-lactam antibiotics. They do this by expressing the β-lactamase gene. When bacterial populations have these resistant subgroups, treatment with β-lactam can result in the resistant strain becoming more prevalent and so, more virulent.

β-Lactams are classified according to their core ring structures:

- β-Lactams fused to saturated five-membered rings;
- β-Lactams containing thiazolidine rings are named penams;
- β-Lactams containing pyrrolidine rings are named carbapenams;
- β-Lactams fused to oxazolidine rings are named oxapenams or clavams;
- β-Lactams fused to unsaturated five-membered rings;
- β-Lactams containing 2,3-dihydrothiazole rings are named penems;
- β-Lactams containing 2,3-dihydro-1H-pyrrole rings are named carbapenems;
- β-Lactams fused to unsaturated, six-membered rings;
- β-Lactams containing 3,6-dihydro-2H-1,3-thiazine rings are named cephems;
- β-Lactams containing 1,2,3,4-tetrahydropyridine rings are named carbacephems;
- β-Lactams containing 3,6-dihydro-2H-1,3-oxazine rings are named oxacephems; and
- β-Lactams not fused to any other ring are named monobactams.

Penicillin (sometimes abbreviated PCN or pen) is a group of antibiotics derived from Penicillium fungi. They include penicillin G, procaine penicillin, benzathine penicillin, and penicillin V. Penicillin antibiotics are historically significant because they are the first drugs that were effective against many previously serious diseases, such as syphilis, and infections caused by staphylococci and streptococci. Penicillins are still widely used today, though many types of bacteria are now resistant. All penicillins are β-lactam antibiotics and are used in the treatment of bacterial infections caused by susceptible, usually Gram-positive, organisms.

The cephalosporins are a class of β-lactam antibiotics originally derived from the fungus Acremonium, which was previously known as "Cephalosporium". Together with cephamycins, they constitute a subgroup of β-lactam antibiotics called cephems. Cephalosporins are indicated for the prophylaxis and treatment of infections caused by bacteria susceptible to this particular form of antibiotic. First-generation cephalosporins are active predominantly against Gram-positive bacteria, and successive generations have increased activity against Gram-negative bacteria (albeit often with reduced activity against Gram-positive organisms).

Antibiotics from Prokaryotes

Most of the currently available antibiotics are produced by prokaryotes mainly by bacteria from the genus Streptomyces. Even though penicillin drugs, antibiotics produced by molds, were the first antibiotics successfully used to treat many serious infections, most of the naturally produced antibiotics are synthesized by bacteria. In 1939 the French microbiologist René Dubos isolated the substance tyrothricin and later showed that it was composed of two substances, gramicidin (20%) and tyrocidine (80%). These were the first antibiotics to be manufactured commercially. Gramicidin is a heterogeneous mixture of six antibiotic compounds, all of which are obtained from the soil bacterial species Bacillus brevis and called collectively gramicidin D.

Streptomyces is the largest antibiotic-producing genus, producing antibacterial, antifungal, and antiparasitic drugs, and also a wide range of other bioactive compounds, such as immunosuppressants. They produce over two-thirds of the clinically useful antibiotics of natural origin. The now uncommonly-used streptomycin takes its name directly from Streptomyces. Aminoglycosides, class of antibiotics, that are derived from bacteria of the Streptomyces genus are named with the suffix -mycin, whereas those that are derived from Micromonospora are named with the suffix -micin. However, this nomenclature system is not specific for aminoglycosides.

Supernatant of a Streptomyces davawensis culture: The picture shows the typical red color of the antibiotic Roseoflavin secreted by the Streptomyces cells in the culture.

Streptomycetes are characterised by a complex secondary metabolism. Almost all of the bioactive compounds produced by Streptomyces are initiated during the time coinciding with the aerial hyphal formation from the substrate mycelium.

Streptomycetes produce numerous antifungal compounds of medicinal importance, including nystatin (from S. noursei), amphotericin B (from S. nodosus), and natamycin (from S. natalensis).

Members of the Streptomyces genus are the source for numerous antibacterial pharmaceutical agents; among the most important of these are: Chloramphenicol (from S. venezuelae), Daptomycin (from S. roseosporus), Fosfomycin (from S. fradiae), Lincomycin (from S. lincolnensis), Neomycin (from S. fradiae), Puromycin (from S. alboniger), Streptomycin (from S. griseus), Tetracycline (from S. rimosus and S. aureofaciens).

Clavulanic acid (from S. clavuligerus) is a drug used in combination with some antibiotics (like amoxicillin) to block and/or weaken some bacterial-resistance mechanisms by irreversible beta-lactamase inhibition.

Other bacterial species produce antibiotics as well. Such an example are some Pseudomonas species which produce antimicrobial compounds. P. aurantiaca produces di-2,4-diacetylfluoroglucylmethane, a compound antibiotically active against Gram-positive organisms. Other Pseudomonas spp. might produce compounds antagonistic to other soil microbes, such as phenazine-type antibiotics or hydrogen cyanide.

Antimycobacterial Antibiotics

Antimycobacterial antibiotics target microbes classified as mycobacterium. Antimycobacterial antibiotics are a class of antimicrobial drugs that target mycobacterium. Mycobacterium is a genus of Actinobacteria that includes pathogens known to cause serious and infectious disease. The types of pathogens considered to be mycobacterium include Mycobacterium tuberculosis (tuberculosis) and Mycobacterium leprae (leprosy). Mycobacterium grow in a mold-like manner on the surface of liquids when cultured. Antiomycobacterial antibiotics specifically target these types of microbes.

A type of antimycobacterial antibiotic includes the class of drugs used for tuberculosis (TB) treatment. The standard "short" course treatment for TB is isoniazid, rifampicin, pyrazinamide and ethambutol for two months, then isoniazid and rifampicin alone for another four months. The patient is considered cured at six months (although there is still a relapse rate of 2 to 3%). For latent tuberculosis, the standard treatment is six to nine months of isoniazid alone.

If the organism is known to be fully sensitive, then it is treated with isoniazid, rifampicin and pyrazinamide for two months, followed by isoniazid and rifampicin for four months. Ethambutol need not be used. Most regimens have an initial high-intensity phase, followed by a continuation phase (also called a consolidation phase or eradication phase) – the high-intensity phase is given first, then the continuation phase.

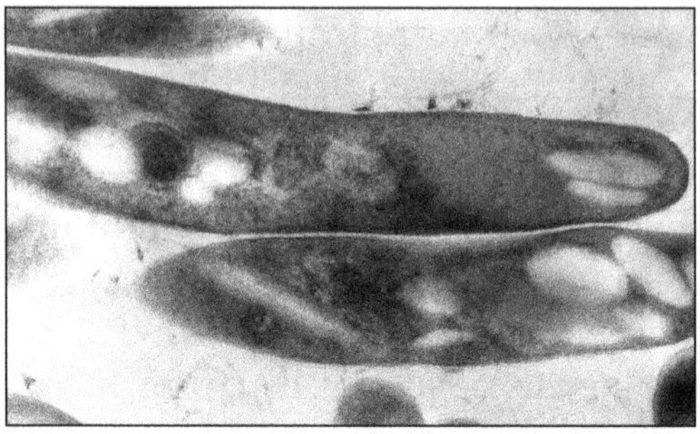

Mycobacterium: Mycobacterium are a class of bacteria defined by their ability to grow in a mold-like manner. Here, a TEM of Mycobacterium tuberculosis, the causative agent of tuberculosis. Antimycobacterial antibiotics target mycobacterium.

There are six classes of second-line drugs (SLDs) used for the treatment of TB. A drug may be classed as second-line instead of first-line for one of three possible reasons: it may be less effective than the first-line drugs (e.g., p-aminosalicylic acid); or, it may have toxic side-effects (e.g., cycloserine); or it may be unavailable in many developing countries (e.g., fluoroquinolones): aminoglycosides: e.g., amikacin (AMK), kanamycin (KM); polypeptides: e.g., capreomycin, viomycin, enviomycin; Fluoroquinolones: e.g., ciprofloxacin (CIP), levofloxacin, moxifloxacin (MXF); thioamides: e.g. ethionamide, prothionamide.

For treatment of leprosy, caused by Mycobacterium leprae, the traditional antimycobacterial drugs include promin (the first treatment introduced to fight leprosy) and dapsone (which eventually become obsolete as Mycobacterium leprae quickly evolved resistance). Modern drugs which were developed in response to the resistance was clofazimine and rifampicin. The use of multidrug therapies including dapsone, clofazimine and rifampicin were advantageous due to the low risk of antibiotic resistance. However, the use of these multidrug treatments was costly and only adopted in endemic countries when the World Health Assembly passed a resolution to eliminate leprosy in 1991.

ANTIMICROBIAL DRUG RESISTANCE

Antimicrobial resistance happens when microorganisms (such as bacteria, fungi, viruses, and parasites) change when they are exposed to antimicrobial drugs (such as antibiotics, antifungals, antivirals, antimalarials, and anthelmintics). Microorganisms that develop antimicrobial resistance are sometimes referred to as “superbugs”.

As a result, the medicines become ineffective and infections persist in the body, increasing the risk of spread to others.

Why is Antimicrobial Resistance a Global Concern?

New resistance mechanisms are emerging and spreading globally, threatening our ability to treat common infectious diseases, resulting in prolonged illness, disability, and death.

Without effective antimicrobials for prevention and treatment of infections, medical procedures such as organ transplantation, cancer chemotherapy, diabetes management and major surgery (for example, caesarean sections or hip replacements) become very high risk.

Antimicrobial resistance increases the cost of health care with lengthier stays in hospitals and more intensive care required.

Antimicrobial resistance is putting the gains of the Millennium Development Goals at risk and endangers achievement of the Sustainable Development Goals.

What Accelerates the Emergence and Spread of Antimicrobial Resistance?

Antimicrobial resistance occurs naturally over time, usually through genetic changes. However, the misuse and overuse of antimicrobials is accelerating this process. In many places, antibiotics are overused and misused in people and animals, and often given without professional oversight. Examples of misuse include when they are taken by people with viral infections like colds and flu, and when they are given as growth promoters in animals or used to prevent diseases in healthy animals.

Antimicrobial resistant-microbes are found in people, animals, food, and the environment (in water, soil and air). They can spread between people and animals, including from food of animal origin, and from person to person. Poor infection control, inadequate sanitary conditions and inappropriate food-handling encourage the spread of antimicrobial resistance.

Resistance in Bacteria

Antibiotic resistance is present in every country. Patients with infections caused by drug-resistant bacteria are at increased risk of worse clinical outcomes and death, and consume more health-care resources than patients infected with non-resistant strains of the same bacteria.

Resistance in Klebsiella pneumoniae – common intestinal bacteria that can cause life-threatening infections – to a last resort treatment (carbapenem antibiotics) has spread to all regions of the world. K. pneumoniae is a major cause of hospital-acquired infections such as pneumonia, bloodstream infections, and infections in newborns and intensive-care unit patients. In some countries, because of resistance, carbapenem antibiotics do not work in more than half of people treated for K. pneumoniae infections.

Resistance in E. coli to one of the most widely used medicines for the treatment of urinary tract infections (fluoroquinolone antibiotics) is very widespread. There are countries in many parts of the world where this treatment is now ineffective in more than half of patients.

WHO recently updated the treatment guidelines for gonorrhoea to address emerging resistance. The new WHO guidelines do not recommend quinolones (a class of antibiotic) for the treatment of gonorrhoea due to widespread high levels of resistance. In addition, treatment guidelines for chlamydial infections and syphilis were also updated.

Resistance to first-line drugs to treat infections caused by Staphlylococcus aureus—a common cause of severe infections in health facilities and the community—is widespread. People with MRSA (methicillin-resistant Staphylococcus aureus) are estimated to be 64% more likely to die than people with a non-resistant form of the infection.

Colistin is the last resort treatment for life-threatening infections caused by Enterobacteriaceae which are resistant to carbapenems. Resistance to colistin has recently been detected in several countries and regions, making infections caused by such bacteria untreatable.

Resistance in Tuberculosis (TB)

WHO estimates that, in 2014, there were about 480 000 new cases of multidrug-resistant tuberculosis (MDR-TB), a form of tuberculosis that is resistant to the 2 most powerful anti-TB drugs. Only about a quarter of these (123 000 cases) were detected and reported. MDR-TB requires treatment courses that are much longer and less effective than those for non-resistant TB. Globally, only half of MDR-TB patients were successfully treated in 2014.

Among new TB cases in 2014, an estimated 3.3% were multidrug-resistant. The proportion is higher among people previously treated for TB, at 20%.

Extensively drug-resistant tuberculosis (XDR-TB), a form of tuberculosis that is resistant to at least 4 of the core anti-TB drugs, has been identified in 105 countries. An estimated 9.7% of people with MDR-TB have XDR-TB.

Resistance in Malaria

Resistance to the first-line treatment for P. falciparum malaria (artemisinin-based combination therapies, also known as ACTs) has been confirmed in 5 countries of the Greater Mekong subregion. In most places, patients with artemisinin-resistant infections recover fully after treatment, provided that they are treated with an ACT containing an effective partner drug. However, along the Cambodia-Thailand border, P. falciparum has become resistant to almost all available antimalarial medicines, making treatment more challenging and requiring close monitoring. There is a real risk that multidrug resistance will soon emerge in other parts of the subregion as well.The spread of resistant strains to other parts of the world could pose a major public health challenge and jeopardize important recent gains in malaria control.

Resistance in HIV

In 2010, an estimated 7% of people starting antiretroviral therapy (ART) in developing countries had drug-resistant HIV. In developed countries, the same figure was 10–20%. Some countries have recently reported levels at or above 15% amongst those starting HIV treatment, and up to 40% among people re-starting treatment. This requires urgent attention.

Increasing levels of resistance have important economic implications as second and third-line regimens are 3 times and 18 times more expensive, respectively, than first-line drugs.

Since September 2015, WHO has recommended that everyone living with HIV start on antiretroviral treatment. Greater use of ART is expected to further increase ART resistance in all regions of the world. To maximize the long-term effectiveness of first-line ART regimens, and to ensure that people are taking the most effective regimen, it is essential to continue monitoring resistance and to minimize its further emergence and spread. In consultation with countries, partners and stakeholders.

Resistance in Influenza

Antiviral drugs are important for treatment of epidemic and pandemic influenza. So far, virtually all influenza A viruses circulating in humans were resistant to one category of antiviral drugs – M2 Inhibitors (amantadine and rimantadine). However, the frequency of resistance to the neuraminidase inhibitor oseltamivir remains low (1-2%).

Need for Coordinated Action

Antimicrobial resistance is a complex problem that affects all of society and is driven by many interconnected factors. Single, isolated interventions have limited impact. Coordinated action is required to minimize the emergence and spread of antimicrobial resistance. All countries need national action plans on AMR.

Greater innovation and investment are required in research and development of new antimicrobial medicines, vaccines, and diagnostic tools.

CAUSES OF ANTIMICROBIAL DRUG RESISTANCE

Microbes, such as bacteria, viruses, fungi, and parasites, are living organisms that evolve over time. Their primary function is to reproduce, thrive, and spread quickly and efficiently. Therefore, microbes adapt to their environments and change in ways that ensure their survival. If something stops their ability to grow, such as an antimicrobial, genetic changes can occur that enable the microbe to survive. There are several ways this happens.

Natural (Biological) Causes

Selective Pressure

In the presence of an antimicrobial, microbes are either killed or, if they carry resistance genes, survive. These survivors will replicate, and their progeny will quickly become the dominant type throughout the microbial population.

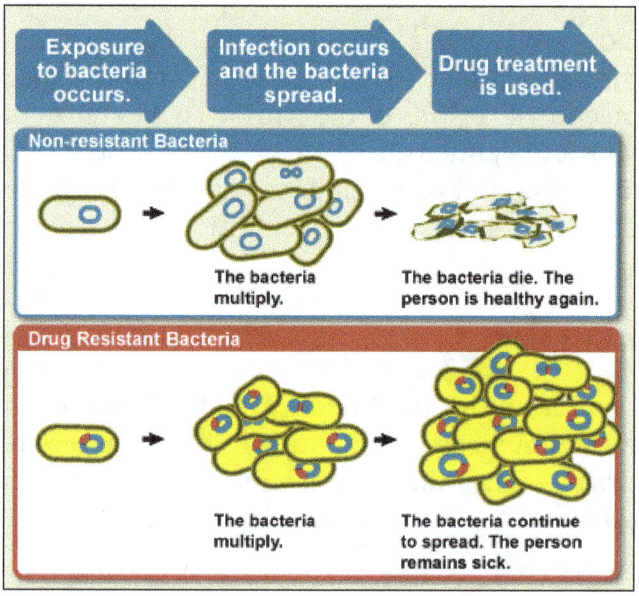

Diagram showing the difference between non-resistant bacteria and drug resistant bacteria. Non-resistant bacteria multiply, and upon drug treatment, the bacteria die. Drug resistant bacteria multiply as well, but upon drug treatment, the bacteria continue to spread.

Mutation

Most microbes reproduce by dividing every few hours, allowing them to evolve rapidly and adapt quickly to new environmental conditions. During replication, mutations arise and some of these mutations may help an individual microbe survive exposure to an antimicrobial.

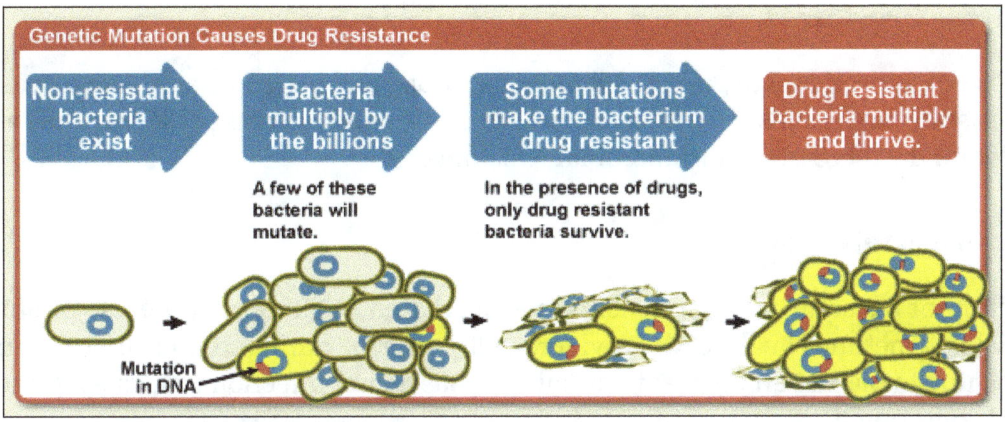

Diagram showing that when bacteria mulitply some will mutate. Some of those mutations can make the bacteria resistance to drug treatment. In the presence of the drugs, only the resistant bacteria survive and then multiply and thrive.

Gene Transfer

Microbes also may get genes from each other, including genes that make the microbe drug resistant. Bacteria multiply by the billions. Bacteria that have drug-resistant DNA may transfer a copy of these genes to other bacteria. Non-resistant bacteria receive the new DNA and become resistant to drugs. In the presence of drugs, only drug-resistant bacteria survive. The drug-resistant bacteria multiply and thrive.

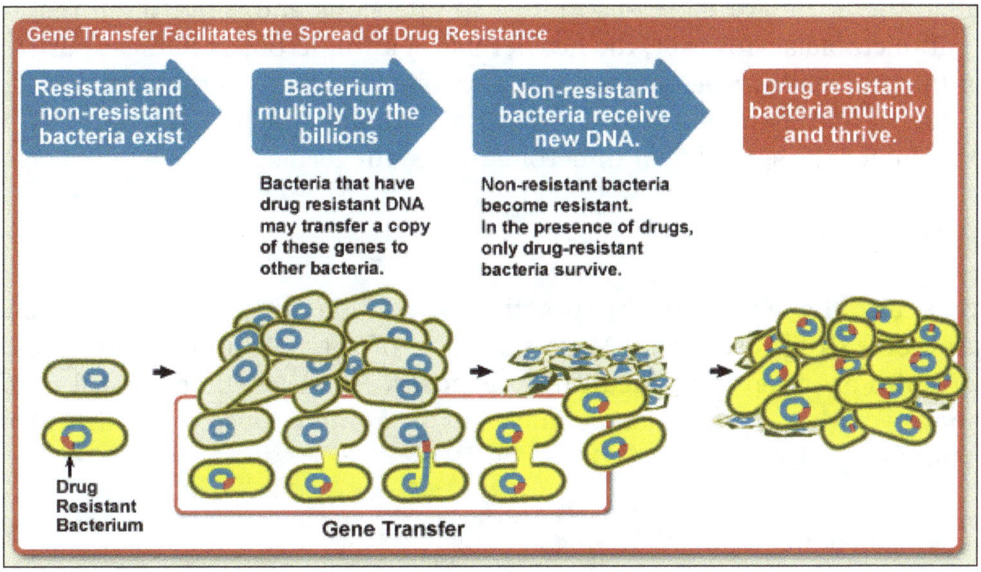

Diagram showing how gene transfer facilitates the spread of drug resistance. Bacteria multiply by the billions. Bacteria that have drug resistant DNA may transfer a copy of these genes to other bacteria. Non-resistant bacteria recieve the new DNA and become resistant to drugs. In the presence of drugs, only drug-resistant bacteria survive. The drug resistant bacteria multiply and thrive.

Societal Pressures

The use of antimicrobials, even when used appropriately, creates a selective pressure for resistant organisms. However, there are additional societal pressures that act to accelerate the increase of antimicrobial resistance.

Inappropriate Use

Selection of resistant microorganisms is exacerbated by inappropriate use of antimicrobials. Sometimes healthcare providers will prescribe antimicrobials inappropriately, wishing to placate an insistent patient who has a viral infection or an as-yet undiagnosed condition.

Inadequate Diagnostics

More often, healthcare providers must use incomplete or imperfect information to diagnose an infection and thus prescribe an antimicrobial just-in-case or prescribe a broad-spectrum antimicrobial when a specific antibiotic might be better. These situations contribute to selective pressure and accelerate antimicrobial resistance.

Hospital Use

Critically ill patients are more susceptible to infections and, thus, often require the aid of antimicrobials. However, the heavier use of antimicrobials in these patients can worsen the problem by selecting for antimicrobial-resistant microorganisms. The extensive use of antimicrobials and close contact among sick patients creates a fertile environment for the spread of antimicrobial-resistant germs.

Agricultural Use

Scientists also believe that the practice of adding antibiotics to agricultural feed promotes drug resistance. More than half of the antibiotics produced in the United States are used for agricultural purposes. However, there is still much debate about whether drug-resistant microbes in animals pose a significant public health burden.

GENETICS OF ANTIMICROBIAL DRUG RESISTANCE

Resistance to an antibacterial agent may be either natural or acquired. Some bacterial species are by nature uniformly resistant, some uniformly susceptible, and some include both susceptible and resistant strains. Natural resistance is reflected by gaps in the spectrum of activity of an antibiotic. Thus, broad spectrum drugs such as tetracyclines are effective against many. bacterial species while narrow spectrum agents such as polymyxins are active against a restricted group of organisms.

Natural resistance (sometimes termed nonsusceptibility) to a particular antibiotic may be caused by a species' impermeability to an agent or by its lack of a target site, which is present in susceptible species.

Acquired resistance develops by mutation or by infection with resistance (R) plasmids. A single mutation may produce a high level of resistance to an antibiotic, such as streptomycin. For other drugs serial passage through gradually increasing concentrations of an antibiotic is required, and the resulting resistant strain generally carries multiple mutations, each providing a small increment in resistance. Resistance resulting from mutation is usually specific for the selecting agent or closely related drugs. It is inherited, but is rarely, if ever, spread to other bacteria. While some resistant mutants retain parental growth and virulence, other mutants are partially crippled. Mutants of this type are likely to be unstable and to revert or be lost due to a disadvantageous growth rate when antibiotic selection is withdrawn. In contrast, acquisition of an R plasmid generally confers resistance to clinically achievable levels of an antibiotic in a single step. A plasmid may carry resistance to one or to many chemically unrelated drugs. Furthermore, plasmids are transmissible by conjugation, transduction, or transformation to other bacteria and can thus disperse their resistance genes. They generally do not have deleterious effects on cell growth or virulence and may in fact carry genes contributing to virulence as well as to antibiotic resistance. Consequently, while study of mutational resistance to antibiotics has revealed much about normal cellular physiology and the actions of antibacterial agents, the major mechanism for resistance in clinical isolates of bacteria is plasmid carriage.

Bacteria with either natural or acquired resistance will be selectively favored in humans, animals, or environments in which antibiotics are used. How commonly this occurs depends on the particular antibiotic and on the genetic potential for development of resistance.

Chromosomally Determined Resistance

Chromosomal mutations conferring resistance to antibiotics occur as spontaneous, random, and relatively rare alterations in the DNA composition of bacteria at frequencies of 10-6 to 10-10 per cell generation. Table lists some well-studied examples of mutational resistance.

Table: Mechanisms of mutational resistance to antibiotics.

Drug	Target	Mechanism of Resistance
Aminoglycosides	Ribosome. Inhibition of protein synthesis	Altered ribosomal proteins (kasugamycin, S2; neamine, S17; streptomycin, S12; spectinomycin, S5)
		Altered ribosomal protein (gentamicin, kanamycin, neomycin, tobramycin, and others, L6)

Drug	Target	Mechanism of Resistance
		Altered 16S RNA methylation (kasugamycin)
		Altered drug transport
Erythromycin	Ribosome. Inhition of protein synthesis	Altered ribosomal proteins (L4, L22)
		Altered assembly of ribosomal subunits
Nalidixic acid	DNA gyrase	Altered subunit of DNA gyrase (nalidixic or oxolinic acid)
		Altered drug transport
Novobiocin	DNA gyrase	Altered subunit of DNA gyrase (coumermycin A_1, novobiocin)
Penicillins	Cell wall biosynthesis	Altered penicillin-binding proteins
		Increased synthesis of chromosomal beta-lactamase
		Altered cell envelope
Rifampin	RNA polymerase	Altered beta subunit of RNA polymerase
sulfonamide	Folic acid biosynthesis	Altered dihydropteroate synthase
Trimethoprim	Folic acid biosynthesis	Thymine auxotrophy sparing folic acid requirement

The biochemical basis of resistance usually involves one of four mechanisms:

- The target site is altered so that binding of the antibiotic is reduced or eliminated,
- There is a block in the transport of the drug into the cell,
- The antibiotic is detoxified or inactivated, or
- The inhibited step in a metabolic pathway is by-passed.

Resistance to the aminoglycosides kasugamycin, neamine, streptomycin, or spectinomycin can result from alteration of the amino acid in a specific protein of the 30S ribosomal subunit. Coresistance to neomycin and kanamycin probably involves the same mechanism but is less well characterized. Ribosomal mutants with broad resistance to aminoglycosides, including gentamicin, kanamycin, neomycin, and tobramycin, also occur, but additional nonribosomal mutations are required to achieve a high level of resistance. Some kasugamycin-resistant mutants have altered 16S RNA methylation.

Lower levels of resistance to aminoglycoside antibiotics can also be attributed to mutations in components of the system by which oxidative energy is coupled to drug transport. Some aminoglycoside transport mutants are broadly resistant to many other related drugs. Resistance to rifampin develops as a result of an alteration in the beta subunit of its target site, RNA polymerase. Escherichia coli mutants that are resistant to erythromycin can be selected with alterations in specific ribosomal proteins or in the assembly of the two ribosomal subunits. Penicillin-resistant mutants can result from alterations of penicillin-binding proteins by increased synthesis of a chromosomally determined beta-lactamase, which is normally produced at low levels or by changes in the cell envelope.

Mutants that are resistant either to nalidixic acid or to novobiocin can result from alterations in components of DNA gyrase activity. Resistance to nalidixic acid can also be caused by altered drug permeability. Finally, resistance to sulfonamide can develop from decreased binding by dihydropteroate synthase, and resistance to trimethoprim can result from mutation to auxotrophy for thymine, thus sparing this important end-product of folic acid activity.

Considering the variety of mechanisms that cause mutational resistance of bacteria to antibiotics, it is perhaps surprising that the genetic mechanism is not more common in clinical isolates. Mutational resistance is clinically important for nalidixic acid, streptomycin, rifampin, and trimethoprim and may be responsible for broadly aminoglycoside-resistant clinical isolates), but for most drugs plasmids provide the major source of resistance to antibiotics.

Plasmids

Plasmids are extrachromosomal genetic elements composed of circular double-stranded DNA. They vary from a molecular weight of approximately 1 million to over 300 million. Hence, they range from one-thousandth to one-tenth the size of the bacterial chromosome and can accommodate from a few to perhaps 500 genes.

Essential genes are concerned with plasmid replication and the partitioning of plasmid DNA to daughter cells at the time of bacterial division. Also in this category are genes that determine the number of plasmid copies per cell chromosome and the plasmid incompatibility behavior. In addition to these requirements, plasmids may carry other genes.

Many plasmids are infectious or conjugative and possess the genetic machinery for assuring their transmission on contact among bacterial cells by conjugation, a complex process that, for one well-studied plasmid, requires at least 17 genes. Plasmids with a molecular weight less than approximately 20 million lack room for this equipment and are generally transfer-deficient (Tra-) although some Tra- plasmids can be transferred ("mobilized") by another conjugative plasmid that is present in the same cell.

A third category of plasmid genes is involved with interactions with other replicons and includes genes that inhibit the propagation of certain bacteriophages or the transfer functions of other plasmids.

Finally, plasmids carry genes that effect the host cell's interaction with the environment. Antibiotic resistance genes are the most familiar, this category also includes genes determining resistance to metallic compounds including arsenicals used as feed additives, genes for resistance to physical and chemical agents that damage DNA, genes for specialized metabolic pathways such as sugar fermentation or other catabolic functions, genes determining bacteriocin production and resistance, and genes involved in pathogenicity such as toxin production or colonization by adherence to mammalian cells.

Plasmids can be detected by physical or genetic techniques. Traditionally, plasmid DNA has most often been isolated by centrifuging a cell lysate in a cesium chloride density gradient containing an intercalating dye such as ethidium bromide that binds differentially to plasmid and chromosomal DNA to allow their separation. Recently, agarose gel electrophoresis has been adapted to allow simpler and more rapid visualization of extrachromosomal DNA. The presence of many plasmids can be demonstrated physically and their molecular weights identified within a day by modifications of this technique. Conjugative plasmids can be detected by testing for transfer of the property they determine to a suitable recipient by mating. Transmission of nonconjugative plasmids can be accomplished by transduction with bacteriophage, transformation with purified plasmid DNA, or mobilization with another plasmid. Finally, certain agents, "curing" compounds, or physical techniques promote the loss of plasmids or select against plasmid-containing cells. Consequently, they can be used to provide presumptive evidence for the presence of a plasmid determined characteristic.

Further physical and genetic characterization of plasmids is generally necessary to identify and assess the spread of plasmids in epidemiological investigations. Basic physical characteristics are plasmid size, DNA composition as expressed by percent guanine plus cytosine, and the fragmentation pattern produced by restriction endonucleases, which are enzymes that recognize specific DNA sequences as cleavage sites. After endonuclease treatment, the fragments that are produced are separated by agarose gel electrophoresis to produce a characteristic pattern for a particular plasmid that is usually, but not always, independent of the host.

Incompatibility behavior is often used to classify plasmids further. Compatible plasmids can coexist in the same host cell while incompatible plasmids cannot and tend to displace each other. By this test plasmids found in enterobacteria can be divided into more than 30 groups. Similar incompatibility schemes allow Pseudomonas plasmids to be classified into 11 or more groups and Staphylococcus plasmids to be assigned to at least seven groups.

Although only a few genes are involved in incompatibility specificity, plasmids belonging to the fame incompatibility (Inc) group usually share much greater DNA homology than do plasmids of different Inc groups, thereby reflecting other similarities in gene function. Plasmids that are transmissible to E. coli have also been broadly divided into those that inhibit the fertility factor (Fi+) and those that do not (Fi-). Bacteriophage interactions can also be used for plasmid classification. Certain phages, termed donor-specific or male phages, adsorb specifically to surface structures, such as thread-like pill, which are part of the conjugation apparatus for transfer-positive (Tra+) plasmids. Other plasmids characteristically interfere with the propagation of certain phages so that both phage susceptibility and resistance can be used for plasmid classification.

Plasmid Transfer

In the laboratory Tra+ plasmids vary in their transfer frequencies from barely detectable values of 10-/to 10-8 to virtually 100% efficiency. Typical average values are 10-3 to 10-5 per donor. The transfer mechanism of some plasmids is naturally repressed, but is relieved from repression when the plasmid enters a new host. Other plasmids transfer much more efficiently on solid rather than in liquid media. For some plasmids transfer is sensitive to temperature, occurring much more readily at 22 C than at 37°C. Such plasmids have been responsible for outbreaks of resistant typhoid fever but could have been overlooked had mating experiments been performed at the conventional temperature.

In-vivo transfer of plasmids appears to be even less efficient than under laboratory conditions, although it has been observed in the gut of both humans and laboratory animals A number of factors have been incriminated. Anderson found that strains containing R plasmids survive less well in the human intestine than do their R- counterparts. The anaerobic environment also reduces the fertility of some plasmids as does the presence of bile salts and the metabolic activity of other gut bacteria, especially Bacteroides. Finally, smooth colony types are less efficient as plasmid recipients than are the rough variants used for laboratory experiments After a volunteer drank a suspension containing 10 donor organisms, Smith found that few ingested R+ Escherichia coli strains were able to colonize the intestine, that transfer to other E. coli occurred rarely, and that these strains persisted for only a few days. In similar studies with even larger inocula, Anderson et al. were unable to detect transfer of R plasmids unless specific antibiotics were also administered, but under these conditions they found that plasmid transfer occurred at high frequency and that R+ organisms could persist for several months. Similar results were obtained by Burton et al. who could only detect transfer of a plasmid determining resistance to tetracycline (among other antibiotics) when 1 g of tetracycline was given daily for 9 days. However, high-level shedding of resistant bacteria persisted from 2 to 43 days in the absence of continued administration of tetracycline. Since plasmid-containing strains, like those sensitive to antibiotics, have variable but finite persistence in the gut flora, continuous administration of antibiotics is more likely to promote plasmid carriage than would intermittent use.

Plasmid Host Range

An important biological limitation to transfer is plasmid host range, which correlates with plasmid Inc specificity. Transfer of intact plasmids between Gram-positive and Gram-negative bacteria has not been accomplished, and transmission of plasmids between facultative and strict Gram negative anaerobes has proven difficult to demonstrate. Many plasmids transfer efficiently to closely related bacteria, for example, among E. coli, Salmonella, and Shigella. Some, such as certain plasmids found in P. aeruginosa, transfer only to other Pseudomonas species. Hence, it is important to use a related organism as a recipient when testing for transmission of a potentially plasmid-determined characteristic.

Other plasmids are much more promiscuous. Members of the IncP group have been transferred to virtually every Gram-negative host tested as recipient. Not surprisingly, P-type plasmids have been found in a variety of natural hosts. Other plasmid Inc groups tend to be associated more often with particular bacterial species, but the number of plasmids examined for host specificity is still small. The question of whether R plasmids of human and animal origin differ has been examined by Anderson et al.. They found a high degree of DNA homology between plasmids of the same Inc group independent of their origin, suggesting a common pool of R factors in humans and animals.

Plasmid Distribution

Plasmidshavebeenfoundinpracticallyallbacteriathathavebeencarefullyinvestigated. Not all are R plasmids. Indeed, some have no established function and are termed cryptic plasmids.

Table lists the antibiotic resistances that have been found on plasmids in particular pathogenic organisms. Gram-negative pathogens have been studied most intensively. R plasmids have been observed in Enterobacteriaceae including Enterobacter, Escherichia, Klebsiellam, Proteus, Providencia, Salmonella, Serratia, and Shigella and in other Gram-negative pathogens including Acinetobacter, Bordetella, Haemophilus, Neisseria gonorrhoeae, Pasteurella, Pseudomonas, Vibrio, and Yersinia. Among the Gram-positive pathogens, plasmids determining resistance are well known in Staphylococcus aureus and S. epidermidis and have also been found in streptococci of groups A, B, and D. Some streptococcal plasmids are even conjugative. Plasmids have been found in S. Pneumoniae as well but have not yet been associated with anti biotic resistance. Recently, R plasmids have also been found in anaerobes including Clostridium perfringens and Bacteroides fragilis. Evidently, plasmids provide a general mechanism for carrying a potentially advantageous but dispensable function in some members of a population in a form that can be spread to other bacteria under selective conditions.

Table: Detection of plasmid-determined antibiotic resistance in pathogens.

Antibiotic										
Organism	Beta-Lactam	Cm	Gm	Km	MLS	Sm	Su	Tc	Tm	Tp
Gram-negative										
Enterobacteriaceae	+	+	+	+		+	+	+	+	+

Haemophilus influenzae	+	+		+				+		
Neisseria gonorrhoeae	+									
Pseudomonas aeruginosa	+	+	+	+		+	+	+	+	
Gram-positive										
Staphylococcus aureus	+	+	+	+	+	+		+	+	
Staphylococcus epidermidis	+	+		+				+	+	
Streptococcus										
Group A				+						
Group B	+			+			+			
Group D	+		+	+	+		+			
Anaerobes										
Clostridium perfringens		+			+			+		
Bacteroides fragilis	+	+			+			+		

Plasmid Prevalence

Studies of primitive societies little exposed to antibiotics and of bacterial isolates from the preantibiotic era indicate that R plasmids preexisted contemporary antibiotic usage. Maré found resistant Gram-negative bacteria in 10% of over 500 fecal specimens from African bushmen and wild animals, but none of the bacteria could transfer their resistance patterns.

The incidence of R factors in intestinal bacteria of healthy people in industrial societies is much higher. For example, Linton et al. reported that 53% of 193 healthy adults and children in Bristol, England, who had not received antibiotics or had not been recently hospitalized, carried antibiotic-resistant coliform bacteria in their feces. Transmissible R plasmids could be demonstrated in 61% of the resistant strains.

Antibiotic usage exerts a strong selective pressure favoring R plasmid carriage. Smith found sensitive E. coli in the fecal flora of 59% of pigs, 73% of chickens, and 90% of calves that had not been fed antibiotics, but in less than 2% of pigs and chickens that had been fed tetracycline. Datta et al. reported that 30% of the women admitted to hospitals in England excreted resistant E. coli but that 18 of 18 given tetracycline had resistant E. coli in their stools and that 66% of these strains contained transmissible R plasmids. Many studies have shown a higher frequency of antibiotic-resistant bacteria among hospitalized patients than among people outside the hospital. There is a higher incidence of antibiotic resistance in coilforms from hospital sewers than from domestic sewers, and multiple resistance is more common.

The incidence of plasmids in hospital isolates is very high. For example, using agarose gel electrophoresis Laufs and Kleimann found plasmid DNA in 76% of E. coli, 31% of Proteus, 95% of Klebsiella pneumoniae, 24% of Pseudomonas aeruginosa, 40% of

Staphylococcus aureus, and 36% of group D streptococci from clinical sources. Multiple plasmids were often present, and 45% of the plasmid-containing enteric strains could transfer antibiotic resistance to an E. coli recipient. In some cases, the presence of multiple plasmids leads to higher levels of antibiotic resistance than is observed with a single plasmid.

In some pathogenic strains multiple resistance is becoming more common. In Japan, where plasmid-determined resistance to chloramphenicol, streptomycin, sulfonamide, and tetracycline appeared in Shigella during the 1950's, this pattern of quadruple resistance has remained by far the most common. In Great Britain, on the other hand, plasmid-determined resistance to streptomycin-and sulfonamide appeared in Salmonella typhimurium in 1963. As antibiotic usage, especially in animal feed, increased, resistance to tetracycline, ampicillin, and kanamycin as well as to streptomycin and sulfonamide became the predominant pattern. In this case the strains acquired additional plasmids. In the last decade, the introduction of newer aminoglycosides has been followed by the appearance and spread of plasmid-determined resistance to amikacin, gentamicin, and tobramycin. In some cases, multiple resistance seems to be acquired as new resistance genes are incorporated into existing plasmids.

Plasmid-determined Antibiotic Resistance

Plasmids determine antibiotic resistance by the same general biochemical strategies but the mechanisms are usually different from those produced by chromosomal mutations. Some specific plasmid-determined mechanisms of resistance, which are listed in table, have been reviewed by Davies and Smith. Penicillins and cephalosporins are hydrolyzed by beta-lactamases, of which 11 types produced by plasmids of Gram-negative bacteria can be differentiated. Chloramphenicol is also detoxified, but by acetylation. The aminoglycosides are attacked by a variety of enzymes that attach acetyl, nucleotidyl (adenylyl), or phosphate groups to key hydroxyl or amino moieties on the antibiotic. Modification of aminoglycoside occurs at a faster rate than the uptake of aminoglycoside so that no active drug is available to block protein synthesis. Tetracycline resistance is not due to drug degradation or modification but rather to a combination of diminished tetracycline uptake and an intracellular inhibition of tetracycline activity.

Table: Mechanisms of plasmid-determined antibiotic resistance.

Drug	Mechanism of Resistance
Aminoglycosides[b]	
Amikacin	6'-N-Acetyltransferase.
Gentamicin	3-N- and 6'-N-Acetyltransferases 2"-o-Adenylyltransferase.
Kanamycin	6'-N-Acetyltransferase 4'-o-Adenylyltransferase 3'-o-Phosphotransferase.

Streptomycin	3”-o-Adenylyltransferase 3”-o-Phosphotransferase.
Spectinomycin	3”-o-Adenylyltransferase.
Tobramycin	3-N- and 6 ‘-N-Acetyltransferase 2”-o-Adenylyltransferase.
Beta-lactams	
Penicillins	Various beta-lactamases (TEM-1, TEM-2, OXA-1, OXA-2, OXA-3, PSE-1, PSE-2, PSE-3, PSE-4, HMS-1, SHV-1.
Chloramphenicol	3-o-Acetyltransferase.
Erythromycin	Methylation of 23S RNA.
Lincomycin	Methylation of 23S RNA.
Sulfonamide	Substitute dihydropteroate synthase.
Tetracycline	Diminished uptake Inhibition of intracellular action.
Trimethoprim	Substitute dihydrofolate reductase.

Plasmid effects on antibiotic uptake have also been suggested for chloramphenicol, kanamycin, and penicillin. Plasmid-determined resistance to the MLS group of drugs (macrolide, lincosamide, and streptogramin B-type antibiotics) involves methylation of adenine residues in 23S RNA of the larger ribosomal subunit so that binding of these drugs to their target is prevented. Resistance to three drugs used in animal feed, oleandomycin, tylosin, and virginiamycin, develops in this way. Finally, resistance to sulfonamides and trimethoprim involves plasmid-determined folic acid biosynthetic enzymes that bypass the metabolic block produced by these chemotherapeutic agents. It is noteworthy that whenever antibiotic degradation is not involved, the drug can remain to exert a continued selective pressure for resistant organisms.

For a few antibiotics plasmid-determined resistance has not yet been found. Specifically, this mechanism of resistance is not known to occur for bacitracin, bambermycin, coilstin, monensin, nalidixic acid, nitrofuran, novobiocin, polymyxin, rifampin, or vancomycin, although resistance to each of these agents can occur intrinsically or by chromosomal mutation.

Plasmid Structure and Rearrangements

For some plasmid systems the resistance and transfer genes occur on separate molecules, each of which is capable of independent replication. Transfer of such an aggregate or multimolecular plasmid is achieved by mobilization of the Tra R-plasmid by its Tra+ partner. In other plasmids the resistance genes and transfer functions are combined in a single molecule which may or may not be able to dissociate into separate replicons. In a cointegrate plasmid the resistance genes are often clustered together on the R-determinant segment of the plasmid, while the genes for transfer and other basic plasmid functions are clustered on the resistance transfer factor (RTF) segment. These two segments may be linked by specific repeated segments of DNA known as insertion sequences since the insertion of such a DNA segment into a structural gene results in its inactivation.

Interactions between paired insertion sequences allow plasmids with this structure to dissociate into two component replicons. The resistance genes may also transpose to another replicon as a unit or duplicate to higher degrees of multiplicity in the presence of an antibiotic to amplify the level of resistance. Individual resistance genes or groups of genes may also be located between repeated segments of DNA, which allow them to be transposed to another replicon such as another plasmid, a phage genome, or the bacterial chromosome. Such a transposable genetic element has been termed a transposon.

Transposons, some of which are listed in table, comprise virtually all types of plasmid-determined antibiotic resistance, including resistance to beta-lactam antibiotics, chloramphenicol, erythromycin, gentamicin, kanamycin, streptomycin, sulfonamide, tetracycline, and trimethoprim. Plasmids must thus be thought of as assemblages of potentially transposable resistance genes. Transposition is not a common event. It generally occurs at frequencies of 10-3 or less, even under the most favorable conditions. However, where the selection pressure is high, the results of transposition events can be dramatic. Multiresistant R plasmids have evolved in this way. Furthermore, a resistance gene that becomes prevalent in one group of organisms, such as the transposon responsible for ampicillin resistance in enteric bacteria, seems to have crossed genetic boundaries to appear in unrelated pathogens and to have caused resistance to penicillin in gonococci and to ampicillin in Haemophilus influenzae by inserting into plasmids that are indigenous to these organisms.

Table: Selected drug-resistance transposons.

Transposon	Drug resistance	Molecular weight ($\times 10^6$)
Tnl	Ampicillin	3.2
Tn4	Ampicillin, streptomycin, sulfonamide	13.6
Tn5	Kanamycin	305
Tn7	Streptomycin, trimethoprim	8.5
Tn9	Chloramphenicol	1.7
Tn10	Tetracycline	6.2
Tn551	Erythromycin, lincomycin	3.5
Tn1696	Chloramphenicol, gentamicin, streptomycin, sulfonamide	9.1

Thus, when one takes into account all the known modes of alteration and transfer of genes in bacteria, it is clear that there is a broad spectrum of factors that can influence the activities and spread of genetic determinants leading to antibiotic resistance. These factors are:

- Mutation: Creation or loss of enzymatic activity by changes in DNA base sequence.
- Derepression: Increase in production of enzyme causing resistance in response to an external stimulus.

- Amplification: Tandem multiplication of R-determinants per plasmid or increased number of plasmid copies per cell.
- Multiple plasmids: Total level of antibiotic resistance sometimes higher than that due to individual plasmids.

Genetic rearrangements:

- Transposition: Insertion of R-determinants from one region of DNA into a new location.
- Fusion: Recombination between two plasmids to produce a composite plasmid.
- Disassociation: Separation of a plasmid into its component replicons.
- Microevolution: Small deletions, insertions, or duplications.

Gene transfer:

- Conjugation: Transfer through cell-to-cell connection of either plasmodia or chromosomal DNA.
- Hormone-stimulated conjugation: Intercellular signalling leading to cell aggregation and high frequency transfer.
- Transduction: Infection by viral particles that carry drug resistance gene(s) from previous host cell.
- Transformation: Uptake of naked DNA from surrounding medium.

Loss of function:

- Genetic deletion.
- Insertional inactivation.
- Plasmid loss or curing.

Plasmids and Virulence

Other genes of importance for bacterial infection are also carried on plasmids. Production of enterotoxin in both Staphylococcus aureus and E. coli may be determined by plasmids. Two types of E. coli enterotoxin are produced: a small, heat-stable polypeptide and a larger, heat-labile protein toxin that has a structure and mechanism of action similar to that of cholera toxin. The production of toxin alone is not sufficient for E. coli to cause disease in animals. Enteropathogenic strains also carry a plasmid that codes for a surface antigen that facilitates adherence to the gut wall.

Species-specific antigens have been described for pig strains (the K88 antigen), calf and lamb strains (the K99 antigen), and strains of human origin. Commonly, enterotoxin-producing E. coli also carry R plasmids, and transfer of resistance often results in concurrent transfer of enterotoxin production. Genes for anti biotic resistance and enterotoxin production can even be linked on the same plasmid, and there is evidence that the genes for toxin production are transposable.

Whether plasmid carriage has other effects on virulence has long been debated. Watanabe initially found smooth R+ strains of Salmonella typhimurium less virulent for mice than their R parents, but he was unable to repeat the observation.

Smith found that a few R plasmids caused a marked reduction in virulence of S. typhimurium for chicks, but that most had only a slight or no effect. The high prevalence of plasmids in hospital isolates of a variety of pathogenic bacteria suggests that most R+ strains are fully virulent. In particular, Brumfitt et al. found that the incidence of resistant fecal E coli was the same in women with or without urinary tract infection, that those women with resistant organisms in the urine carried the same organism in the stool, and that R plasmid carriage neither hindered nor facilitated infection of the urinary tract by intestinal E. coli.

E. coli that cause extraintestinal infections in humans are not likely to be toxigenic or invasive. Whether they have other virulence properties is a subject of active current investigation. Smith found that an E. coli strain causing bacteremia in a lamb carried a virulence plasmid (Vir) that determined an exotoxin, which was lethal to other animals when intravenously injected. Only one of 190 E. coli isolates had the same property and it also came from a lamb with bacteremia.

In the same study, Smith also found an E. coli strain isolated from an outbreak of bacteremia in chickens that could transfer increased animal lethality to other E. coli. Increased survival in blood and peritoneal fluids rather than toxin production was associated with this virulence factor, which proved to be identical with the Col V plasmid. The prevalence of Col V is high in invasive E. coli strains from animals and humans. The production of colicin per se is not essential for enhanced virulence. Rather, increased virulence correlates with serum resistance and, hence, with decreased sensitivity to host defense mechanisms that depend on antibody and complement. Col V plasmids can carry antibiotic resistance genes. Therefore, as in the case of enterotoxin production, factors that increase the prevalence of R plasmids can also increase the prevalence of virulence factors.

Attack on R Factor Replication or Transfer

A wide variety of physical and chemical agents, including some antibiotics and at least one drug used in animal feed, bambermycin, can block plasmid transfer or enhance loss in vitro. Conjugation, which is probably the predominant mode of plasmid transfer in nature, requires several conditions. The formation of sex pill, which is essential for mating pair formation, requires a certain temperature range. For the well studied F factor, this range is approximately 30° C to 43° C. The recipient cell surface must also contain certain components such as outer membrane proteins, the lack of which results in conjugative deficiency. In the laboratory, plasmid transfer can be blocked by the addition of a small amount of sodium dodecylsulfate, which presumably disrupts either the recipient or donor cell surface. Rifampin, levallorphan, phenylethyl alcohol, donor-specific phages, or the concomitant presence of an Fi+ plasmid prevent formation or function of sex pill as do mutations in the plasmid genes that confer transmissibility (tra genes). Nalidixic acid also blocks DNA transfer during conjugation, presumably by its action on DNA gyrase.

In many studies these agents have been tested against only a limited number of plasmids. Consequently, whether their effects apply to all plasmids is not known. Their effectiveness in blocking transfer in vivo has not been studied. It is possible that mutants resistant to their action could develop. Further studies are needed to determine the potential value of this approach.

IMPORTANCE OF ANTIMICROBIAL DRUG RESISTANCE MONITORING

Antimicrobial resistance (AMR) is an ongoing problem with multidrug-resistant strains of bacteria and fungi impacting medical progress in many regions of the world. Collecting AMR surveillance data is an essential approach to (1) define the scope of the resistance problem, (2) develop interventions that improve the appropriate application of antimicrobial agents, and (3) decrease resistance selection pressure. Other important efforts are underway to understand the mechanisms of resistance whereby microorganisms avoid the effects of antimicrobials and to use that information to discover/develop new compounds, or modify older agents, that retain potent activity against key target pathogens. Trends in AMR have been described in numerous single-center and population-based surveys conducted throughout the world. However, the dynamic nature of AMR trends in the United States and elsewhere suggests that this issue still merits considerable monitoring and more comprehensive, well organized surveillance programs.

One important aspect of any antimicrobial surveillance program is longitudinality. By conducting surveillance of specific pathogens over time, one can assess the emergence of specific strains or species and discover changes in the antimicrobial susceptibility profile of the organisms. Furthermore, when longitudinal surveillance encompasses a broad geographic distribution, one may eventually develop a useful understanding of regional, national, or even global trends of species distributions and AMR.

The SENTRY Antimicrobial Surveillance Program (SENTRY Program) was designed to track AMR trends and the spectrum of microbial pathogens across various infection types on a global scale. The SENTRY Program has unique features that distinguish it from other excellent contemporary surveillance designs. Whereas most surveillance initiatives are based in a single nation, may track only nosocomial infections, and/or rely primarily on categorical susceptibility testing result inputs from participating centers (diverse susceptibility testing methods), the SENTRY Program monitors both nosocomial and community-onset infections on a global scale and uses validated reference identification and susceptibility testing methods via a central monitoring laboratory model.

The SENTRY Program, which is being recognized for its 20 years of existence through this publication, originated from the recommendations of the American Society for Microbiology (ASM) Task Force on Antimicrobial Resistance that convened in 1994. That task force was assembled to discuss the concern of rising AMR rates and the lack of available antimicrobial agents to treat emerging health threats. Three major components were outlined during that meeting: (1) education, (2) basic research and development, and surveillance. The surveillance component highlighted a number of variables that notably included such an endeavor and would likely require a consortium funding approach that would involve government, professional societies, independent foundations, and pharmaceutical industry contributions. Unfortunately, it was soon recognized that no mechanism existed to bring these stakeholders together in a complementary manner to launch a jointly funded national surveillance system.

Shortly after the task force meeting, Professor Ronald Jones, MD (University of Iowa College of Medicine) was approached by Bristol-Myers Squibb to implement a surveillance program in which its portfolio of antimicrobial agents and other marketed therapeutic drugs would be monitored against a large and diverse collection of clinical isolates across years. This initial concept morphed into the SENTRY Program, which was launched in 1997.

The SENTRY Program utilized many of the components desired by the ASM task force, including a focus on monitoring the most prevalent bacterial and fungal pathogens in contemporary human infectious diseases through various objectives, a wide net cast globally to recruit and collect organisms from a large number of sentinel medical centers (hence, the name SENTRY) that represented a broad geographic human population, and a central laboratory processing approach to apply standardized reference antimicrobial susceptibility testing methods. The study was jointly coordinated through the University of Iowa, University of Utrecht and the Women's and Children's Hospital and regional testing was performed at those 3 reference laboratories. Scientific oversight was initially provided through Professor Jones and a consortium of key opinion leaders.

One important distinction of the SENTRY Program was applying prevalence-based collection objectives in the surveillance model whereby consecutive isolates per infection type were collected overall and/or on a month-to-month basis to diversify the submitted isolates and reduce seasonality or other epidemiological concerns related to specific pathogens. At the onset of the program, 6 core collection objectives addressed the most common types of infection: bloodstream infections, community-acquired respiratory tract infections (fastidious pathogens only), pneumonias in hospitalized patients, skin and soft tissue or wound infections, urinary tract infections, and invasive fungal infections. Although the total contribution of clinical isolates per objective has varied over the years, these major collection objectives remain intact today. Other collection objectives have been introduced throughout the life of the SENTRY Program, and some of these represented a one-time objective for scientific pursuits (eg, gastroenteritis pathogens) or were introduced into the surveillance life cycle due to the emerging needs of drug development and indications being sought for regional regulatory submission (eg, intra-abdominal infections).

Another important aspect of SENTRY Program surveillance is the breadth of antimicrobial agents included in the program each surveillance year. The genesis of our program support was provided by a single pharmaceutical sponsor, and investigational agents were initially limited to that sponsor's product, in addition to other relevant and clinically available agents used for contemporary infectious disease practice. Eventually, the SENTRY Program moved to the consortium-funded design approach in which numerous sponsors funded the program through contractual support. This model allowed numerous investigational drugs and combination agents to be included in the surveillance and each isolate tested to have 20–30 drugs included in the antibiogram profile. Contributing SENTRY Program medical centers can access their antibiogram data (for approved drugs) for submitted isolates, providing an opportunity to gain immediate access to newly approved drug susceptibility information that may not be accessible by other means due to limited diagnostics. It has become clear that automated susceptibility testing system approvals for new drugs can significantly lag behind US Food and Drug Administration release.

Other notable SENTRY Program features include the central reference laboratory design, where reference testing reagents and standard operating protocols and methods are used in the yearlong testing schedules that include rigid quality assurance. In addition, where possible, the same sites have been recruited and included each year to preserve the longitudinality of surveillance data.

Expanded technology has also played an important role in the SENTRY Program. Several advancements have improved the quality and speed of the testing services and analyses performed. Initial organism identification methods and processes relied on traditional microbiologic methods and training, along with biochemicals and other automated commercial testing systems (Vitek, and Vitek 2; bioMerieux). Eventually, matrix-assisted laser desorption ionization-time of flight mass spectrometry (MALDI-TOF MS) became the preferred technology for identification due to its rapid, accurate results and low cost per test. However, there are still many organisms for which traditional manual methods and/or biochemical and other reagents are used to differentiate species or provide backup for MALDI-TOF MS in cases in which the confidence score does not meet the threshold for certain identification confirmatory assignment. Sequencing methods are also applied for identification purposes, as needed.

Initial epidemiological and molecular screening for genetic relatedness and resistance mechanisms were performed using riboprinting and traditional polymerase chain reaction methods. Eventually, the progression of resistance among Gram-negative species isolates led to the growth in deoxyribonucleic acid microarray techniques and methods for the rapid detection of β-lactamase enzymes and other important resistance mechanisms. Gene sequencing was always viewed as an important and vital tool in describing the genome of unique and resistant pathogens, but at the outset of the SENTRY Program and only until recently, a large-scale investigation using sequencing methods was not a scalable option due to the costs and relevant timing. Within the last few years, advances in next-generation sequencing (NGS) equipment, both in terms of acquisition and testing costs, and small equipment footprint designs have made it routinely possible to apply NGS methods for our molecular investigations.

Finally, advances in computers, software design, and capabilities including available laboratory information management systems (LIMS) and web-enabled program features and accessibility have greatly improved the operational and data analysis aspects of the SENTRY Program surveillance. Initial program oversight and data review were predominantly performed manually and individual isolate reviews were completed by laboratory scientists. Internally and externally developed LIMS programs have allowed for better management of the operational process. Computer-based algorithms led to a rapid and more efficient review of data, resulting in high-quality assurance of testing results and recognition of outliers (nonwild-type strains or potential test errors) in drug-pathogen results. Laboratory information management systems also created greater efficiencies that allowed for higher volumes of tests to be performed, including the expansion of antimicrobial agents tested per pathogen processed.

ANTIBIOTIC RESISTANCE

Antibiotic resistance is the ability of bacteria or other microbes to resist the effects of an antibiotic. Antibiotics, representing both naturally as well as chemically synthesized entities, emerged as a powerful tool in counteracting infectious diseases, following serendipitous discovery of penicillin from Penicillium notatum by Alexander Fleming in 1928. Widespread usage of antibiotics that imposes strong selection pressure for resistance development (ability to withstand effects of antibiotics) took a strong grip over the health care system globally as concerns regarding resistance to available drug regime restrict therapeutic options available to treat the disease. Emergence of resistance at rapid pace made the pathogens well-fit and well-adapted, resulting in causing serious life threatening complications as we lack robust drugs to curb the menace of multidrug resistance. Growing menace of antibiotic resistance is inevitable fallout of the introduction of new antibiotics aimed at long-term efficacy in the treatment of infectious diseases. Deteriorating public health ensuing emergence among pathogenic and commensal bacteria of resistance, illustrates a grave predicament globally. Steadily increase in the development of resistance among bacteria thwarts current treatment regimes in hospitals and community settings. Through each passing day, treatments of infectious diseases require administration of high doses of antibiotics and longer stay in hospital. Widening gap between lean productions of drugs increases need of either to rejuvenate the drying antibacterial pipelines or design innovative strategies to combat bacterial antibiotic resistance.

Antibiotics and Bacterial Resistance

Antibiotics are the agents used commonly in the treatment and prevention of infections. Owing to their structure and degree of affinity to target sites, they are classified into Penicillin's, Cephalosporins, Tetracyclines, Aminoglycosides, Macrolides, Sulfonamides, Quinolones, Diaminopyrimidines, Polymyxin and Carbapenems. Being specific in their effect toward different bacterial species, antibiotics culminates them either by: (i) affecting cell wall synthesis (β-lactams), (ii) by targeting protein synthesis machinery via, interaction with ribosomal subunits (Tetracycline, Chloromphenicol, Aminoglycosides etc), (iii) disrupting with nucleic acid machinery (Rifampcin, Fluoroquinolones), (iv) interfering with metabolic pathways (Folic acid analogs, sulfonamides), and (v) by disrupting bacterial membrane structure.

Resistance Mechanisms

Aminoglycosides

The main aminoglycoside resistance mechanism involves modification of the enzymes. Three major classes of proteins are classified in accordance with the kind of modification: AAC (acetyltransferases) which are AAC(1), AAC(2), AAC(3), and AAC(6); ANT (nucleotidyl transferases or adenyl transferases), which includes five nucleotidyl transferases: ANT(2), ANT(3), ANT(4), ANT(6), and ANT(9), and APH (phosphotransferases) which includes seven phosphotransferases: APH(2), APH(3), APH(3), APH(4), APH(6), APH(7), and APH(9).

β-Lactam

Resistance is acquired through production of β-lactamases, like extended-spectrum β-lactamases (ESBLs), ESBL genes (blaCTX-M, blaSHV, blaTEM) plasmid-mediated

AmpC enzymes, and carbapenem-hydrolyzing β-lactamases (carbapenemases). Though, Stenotrophomonas maltophilia have endogenous metallo β-lactamases (MBL) L1 that makes it resistant to carbapenems. Carbapenem resistance among gram positive bacteria is acquired by mutations in the penicillin binding proteins (PBPs). However, in gram negative bacteria, lower penetration of the drug through decrease in the expression of outer membrane porin proteins such as OprD of Pseudomonas aeruginosa. A tripartite efflux pump that causes exclusion of carbapenems from periplasmic space, adds to carbapenems resistance. Additionally, carbapenemases also contributes to carbapenem resistance. The main efficient carbapenemases responsible for carbapenem hydrolysis and its geographical dissemination are KPC, VIM, IMP, NDM, and OXA-48 types. In a plasmid of K. pneumoniae HS11286 strain it was seen that deletion of bla KPC−2 abolished resistance toward carbapenem (cefoxitin, ceftazidime), and exhibited dose-dependent susceptibility toward cefepime supporting that bla KPC−2 is a key factor for the resistance toward cephalosporins and carbapenems in K. pneumoniae.

Chloramphenicol

It acts as a broad spectrum antibiotic resistance mechanism for chloramphenicol involves enzymatic inactivation via acetylation mediated by chloramphenicol acetyltransferases (CATs). Apart from enzyme inactivation chloramphenicol resistance mechanisms, also involves inactivation by phosphotransferases, target site mutation, permeability barriers and efflux pumps.

Glycopeptide

The vancomycin resistance originated from the production of modified peptidoglycan precursor, d-Ala–d-Lac (VanA, VanB, and VanD) or d-Ala–d-Ser (VanC, VanE, and VanG), to which glycopeptides display diminished binding affinities. The vanA and vanB operons are positioned on plasmids as well as on chromosome; whereas the vanC1, vanC2/3, vanD, vanE, and vanG solely show their presence on chromosomes.

Quinolone

Though resistance mechanism for quinolone was found restricted to chromosomes, three plasmid-mediated resistance mechanisms have also been reported. The chromosome-encoded resistance produce a declined outer-membrane permeability linked with porin loss, while over expression of the naturally existing efflux pumps create mutations in the molecular targets, DNA gyrase and topoisomerase IV. Mutations were found occurring at quinolone resistance determining regions (QRDR) in the genes gyrA, gyrB, parC, and parE; which program the subunits of DNA gyrase and topoisomerase IV. Despite the fact that qnr determinant is the first recognized plasmid-mediated quinolone resistance gene, five new lineage of qnr genes have been accounted: qnrA, qnrB, qnrC, qnrD, and qnrS. Second kind of plasmid positioned quinolone resistant gene is a cr variant of aac(6)-Ib, that is aac(6)-Ib-cr, encoding aminoglycoside acetyl transferase. The third means of resistance involves qepA, a plasmid-mediated efflux pump along with its E. coli derivative QepA2, is able to expel hydrophilic fluoroquinolones, e.g., ciprofloxacin.

Sulfonamide

Sulfonamide resistance in chromosome appears through mutations in the folP gene, encoding dihydropteroate synthase. Acquired sulfonamide resistance was identified in the 1960s, and the plasmid-mediated genes sul1 and sul2 were described after 1980s. In addition a third plamid mediated gene sul3 has also been recognized.

Tetracycline

Mechanisms of resistance for tetracycline hold three key strategies: energy-dependent efflux pumps (ABC efflux pumps) ribosomal protection proteins {RPPs, Tet(O)} or enzymatic inactivation.

Colistin

Modifications in the lpxA, lpxC, and lpxD genes of A. baumannii result in neutralization of lipid A biosynthesis, causing total loss of LPS leading to a loss of the polymyxin target. Polymyxin resistance is controlled by two-component systems PhoP/PhoQ and PmrA/PmrB, which react to cation (calcium, iron, and magnesium) concentrations and pH variations. These systems are concerned in the alterations of LPS resulting in polymyxin resistance. Several molecular mechanisms have been associated with colistin resistance in Gram-negative bacteria, like modifications with PmrA/PmrB, PhoP/PhoQ, ParR/ParS, ColR/ColS, and CprR/CprS two-component systems and alterations in the mgrB gene, that codes for negative regulator of PhoPQ. Addition of cationic groups on lipid A due to mutations creat less anionic lipid A ultimately causing less fixation of polymyxins. The polymyxins remains one of the last classes of antibiotics in which resistance is not known to spread from cell to cell via plasmid mediated. There is a current report of plasmid mediated colistin resistance in china designated as mcr-1 gene which is also reported closely in five continents viz, Asia, Europe, Africa, North America, and South America.

Antibiotics do not, in themselves, cause resistance but frequent and high exposure of antibiotics to bacteria creates a selection pressure which triggers resistance strategies of bacteria. Acquirement of resistance genes has been viewed as main donor in favor of the extensive dispersal and increase in antimicrobial resistance through horizontal transfer involving MGEs. Their presence on mobile genetic elements facilitate transfer to un-related bacteria in a process referred to as horizontal gene transfer (HGT) via, conjugation, transduction, or transformation. Transformation involves movement of cellular DNA among closely linked bacteria, imparted by chromosomal set of proteins that occur in naturally transformable bacteria. Conjugation needs autonomously replicating genetic elements known as conjugative plasmids that cause movement of plasmid from the donor cell to a recipient cell that is devoid of it. Transduction involves transfer of DNA facilitated by bacteriophages, constituted host DNA in their capsid and insert this DNA into a new host, where it combines with cellular chromosome and is inherited. Movement of genes confers new metabolic capabilities to the recipient, thereby helps them in their adaptation to new ecological niches. Resistance to antibiotics conferred by chromosomal or mobile genetic elements, is achieved by following strategies: (i) reduction of membrane permeability to antibiotics either by decreasing uptake or increasing efflux, (ii) drug inactivation either by hydrolysis or by modification, (iii) alteration in drug target and decreased binding permeability, and (iv) mutation.

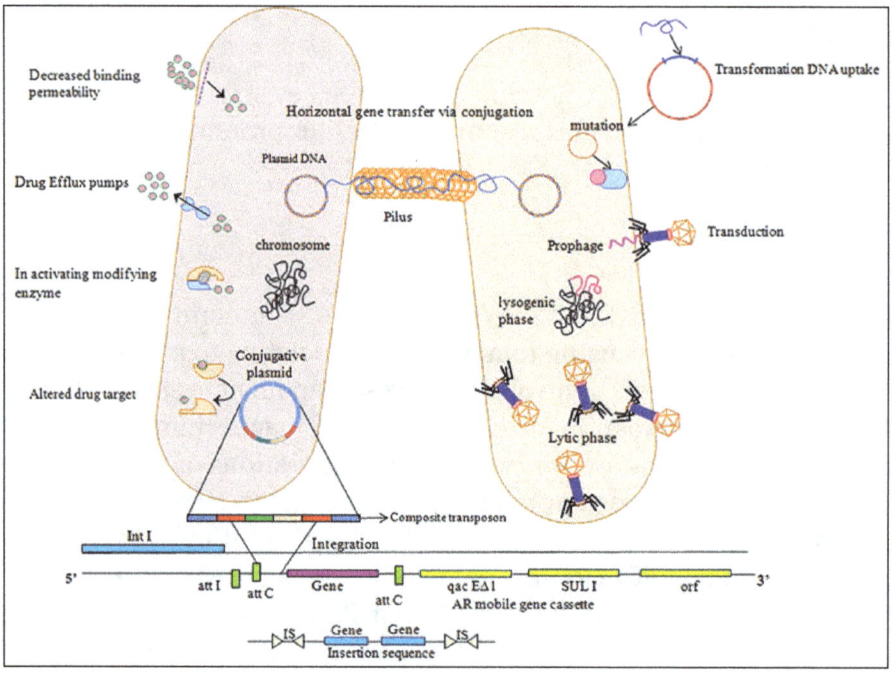

Various ways of resistance mechanisms to counteract effect of antibiotics. Horizontal Gene transfer facilitates transfer and exchange of genetic material among bacterial cells. Transformation involves direct uptake of genetic material from the surrounding by competent recipient having chromosomal set of proteins. Transduction involves DNA insertion into chromosome as prophage which then replicates, packages host DNA alone or in combination with the host cell chromosome. Conjugative plasmids utilize a protein structure pilus to make a link with the recipient cell so as to move them into the recipient cell that ultimately transfers the copy of entire bacterial chromosome, multicopy plasmid or a small portion to a recipient cell, where these genetic elements insert into the chromosome or replicate independently if compatible with the inhabitant plasmids. Integrons use site specific recombination mechanism where it provides a promoter for gene cassettes to exchange and disseminate. Transposons and insertion sequences insert into new sites on the chromosome or plasmids by non-homologous recombination and increase the copy number of transferred genes giving rise to chromosomal mutations, deletions and rearrangements.

Genetic basis of Antibiotic Resistance

Bacteria appeared on this planet billion of years ago, so have their skills sharpened due to genomic flexibility at shielding themselves from toxic chemicals. Bacteria are well-known potent originators for the dissemination of antimicrobial-resistant genetic apparatus. They are competent to offer secured platform for the upholding and transmission of genes accountable for antimicrobial resistance as part of mobile genetic elements (MGE; plasmids, transposons and integrons). The transposons and integrons, owing to their genomic plasticity have contributed a great deal to the fitness quotient and robustness of bacteria to survive in varying environments. Integrons, typically transported by plasmids or enclosed in transposons, performing the task of resistance gene dissemination plays an important role in the revealing of Super Bugs. Since its earliest assessment in 1989 molecular mechanisms involved in the mobility of integrons, their excision and integration for gene cassettes, is currently being scrutinized. Establishing role of MGEs in genomic evolution justifies the predictions of Barbara McClintok that transposons play a major role in the genomic diversity and evolution. Owing to their capacity to relocate between host genomes, MGEs play a vital function of acting as vehicles for resistance gene acquisition and their successive propagation.

Resistance Mediated by Plasmids

Plasmids that mediate horizontal movement of plasmid-borne genes are accountable for global spread of resistance. Resistance plasmids (attributing resistance to commonly used antibiotics) are mostly conjugative; additional are mobilizable. Conjugative plasmids display both broad (no host restriction within the division) and narrow (shifting limited to small number of related bacterial groups) host range. Genes acquired through homologous recombination, integration and excision from the host chromosome relocate from donor to recipient cells by conjugation. These type of plasmid-encoded complexes help the contributor by attaching to promissing recipient that lead to the generation of secured association, required prior to the relocation of DNA. Plasmids that fails to get reloctaed by this approach are transferred by conjugative elements subsequent to the development of transitory or steady fusions called cointegrates. Plasmids also encourage cell contact development through production of pheromone influenced microfibrillar exterior covering substances. Mobilizable plasmids carry DNA transfer genes essential for structure of all or element of the relaxasome, but are deficient of genes essential for mating pore formation. They have a capacity to use conjugative plasmids for horizontal spread; these are immobile in cells which are short of mobile elements carrying compatible mating-pore genes. The majority of known mobilizable plasmids utilize conjugative element mating-pores by expressing their own relaxase (Mob) which works on the plasmid's cognate oriT. Current studies have revealed that plasmid transfer can also take place even when the mobilizing plasmid and the plasmid being mobilized are in two diverse bacterial cells. This type of recruitment, in which a contributor strain possessing a self-transmissible plasmid is getting a second plasmid from a receiver strain, is known as retrotransfer. Retrotransfer take place through two stages; (1) self-transmissible plasmid move from the contributor to the receiver, and (2) mobilization of plasmid from receiver back to the contributor. As the capacity of a self-transmissible plasmid to promote the acquirement of novel plasmids by its bacterial host possibly grant a benefit to the contributor bacterium, it can be said retrotransfer may play a significant job in the progression of plasmid transfer system.

Transphylum mobilization events that incorporate elements from entirely diverse phylogenetic group of bacteria, underscores broad range of interactive capacity originated in gene relocation elements. Besides being recognized through PCR amplification of identified incompatibility groups, mobilizing plasmids are identified by conducting a triparental mating among E. coli having a mobilizable IncQ plasmid, a recipient lacking plasmid and amalgam of local soil or marine bacteria. Endurance of plasmids in natural isolates and their perceptible firmness in absence of antibiotics, opposes a common thought that in non-existence of selection pressure plasmids additional gene transport factors are easily lost. The multitude of antibiotic-resistant strains in environmental milieu where bacteria apparently do not appear to be in touch with antibiotics, suggest that resistance genes can also be firmly retained even in the paucity of antibiotic selection. As plasmids exhibit remarkable property of crossing species borders effortlessly co-mobilization of resistance genes aggravates furthermore clinical crisis.

Resistance Mediated by Transposons

Transposable elements (TE) are the DNA sequences that provide flexibility to the genome. Being proficient to alter their position, they are able to alter their genetic background along with that they change the genetic setting of the locus, where they get inserted. Based on their role in identification and recombination of particular sequences, TEs are categorized into two classes; composite transposons (Class I; holding a range of resistance genes which possess identical structural and functional characteristics, but small DNA homology) and complex transposons (class II; constituting three dissimilar but interrelated families; Tn3, Tn21 and Tn2501). Some of the composite transposons in gram negative bacteria are Tn5, Tn9, Tn10, Tn903, Tn1525, and Tn2350 and among gram positive bacteria are Tn4001 and Tn4003. Compared to Tn1, Tn3, Tn21, Tn501, Tn1721, and Tn3926 found among gram negative bacteria, gram positive bacteria encompass Tn551, Tn917 and Tn4451 complex resistance transposons. These components are possessed with the capability to progress both intra and inter-molecularly which means they can jump within a DNA molecule or from one DNA molecule to another. Tn21 being majorly studied, bear OXA (a carbapenems, possess oxacillinase activity) and PSE (β-lactam gene Pseudomonas specific enzyme) determinants that makes them resistant to aminoglycoside antibiotics. Tn21 also show resistance toward mercury compounds and trimethoprim imparted by dhfr II and V. Class I or retro-transposons work by copying RNA from DNA by transcription and

RNA to DNA by reverse transcription; thereby get inserted into the genome at a diverse location. Acting in cut and paste manner, class II transposons does not involve RNA intermediate. These, transposases create staggered cut at specific site, creating sticky ends; following its transposition to the aimed site, generally followed by target site duplication and construction of short direct repeat at insertion sites. Though transposons provide antibiotic resistance due to the existence of an extra gene on a plasmid, there are chances that transposons can jump from chromosomal DNA to plasmid DNA and vice versa for development of resistance.

Insertion Sequences (ISs; size < 2.5 kb) are basic form of mobile genetic elements disseminated in bacteria. ISs are contemplated as non-complex bacterial mobile DNA taking into account their structure. They include more than 19 families, having dissimilar size. ISs include an open reading frame that codes for a transposase enzyme, surrounded by inverted repeat sequences of 10–40 base pairs at both ends. The transposase enzyme cuts target DNA and inserts the IS due to possible association with the inverted repeat sequences. Exhibiting fondness toward AT-rich region of DNA, higher chances of undergoing homologous recombination, creates variety of possibilities such as deletions, inversions and duplications. There are evidences that when two identical IS elements surround a region of DNA, a composite transposon is produced, and the total interceded DNA flanked by the terminal inverted repeats get mobilized by one or both of the IS coded transposases.

Resistance Mediated by Integrons

Integrons attribute a great deal to the fitness quotient and robustness of bacteria to survive in varying environments. Harboring resistance determinants such as antibiotic resistance genes, their mobilization as part of chromosomes and plasmids and integration far off from their origin confer resistance to antimicrobials. Their categorization is based on amino acid sequences of integrase IntI; those carrying IntI1 are referred as class 1, IntI2 as class 2, IntI3 as class 3 and so on. Integrase IntI1, IntI2, and IntI3 were found associated with mobile genetic elements, while IntI4 was found linked with chromosomal integrons.

Class 1 Integrons

Class I integrons are found associated with the acquisition and mobilization of antibiotic resistance genes. Originated from Tn402, they are composed of two sequence; 5′ conserved sequence (5′CS) representing an integrase gene and a 3′ conserved sequence (3′CS) encoding quaternary ammonium compound resistance gene (qacΔE1) and sulfonamide resistance gene (sul1), respectively. With three recombination sites (attI1, attC and secondary site), expression of captured gene cassettes acquired via site-specific recombination is driven by a promoter located in the 5′-conserved segment (5′-CS) region. Class 1 integrons are associated with a variety of resistance gene cassettes, but most integrons contain an aadA resistance determinant, encoding streptomycin-spectinomycin resistance. Trimethoprim resistance determinants are also detected frequently. Showing prevalence of 22–59%, its localization is reported among diverse groups of Gram negative bacteria; Escherichia, Klebsiella, Aeromonas, Enterobacter, Providencia, Mycobacterium, Burkholderia, Alcaligenes, Campylobacter, Citrobacter, Stenotrophomonas, Acinetobacter, Pseudomonas, Salmonella, Serratia, Vibrio, and Shigella. Gram positive bacteria; Enterococcus, Corynebacterium, Streptococcus, Brevibacterium, Aerococcus, and Staphylococcus show high prevalence of aadA and dfrA gene cassettes.

Class 2 Integrons

Class 2 integrons associated with the Tn7 transposon family (Tn1825, Tn1826, and Tn4132), carry a recombination site attI2 and promoter within these transposons. Its 3′ conserved segment (3′-CS) contains 5 tns genes (tnsA, tnsB, tnsC, tnsD and tnsE) associated with movement and preferential insertion at unique site within bacterial chromosomes. The amino-acid sequences coded by intI2 gene show <50% homology with intI1, and its non-functionality was found attributed by replacement of glutamic acid with a termination codon that leads to production of a shorter and inactive polypeptide. The classic structure of class 2 integrons contain a range of gene cassettes, including streptothricin acetyltransferase (sat1), adenyltransferase (aadA1), dihydrofolate reductase. Class 2 integrons have been reported among Salmonella, Enterobacteriaceae, Acinetobacter, and Psuedomonas.

Class 3 Integrons

Class 3 integrons (IntI3) create excision of integrated cassettes and integration of circularized cassettes into the attI3 site with a considerably lower recombination than that observed with IntI1. This class of integron was firstly identified from Serratia marcescens in 1993, and then found associated with blaGES-1 from Klebsiella pneumoniae strain FFUL 22K. Class 3 integron containing blaGES-1 within the IncQ plasmid was also found in E. coli. Occurence of class 3 integrons associated with IMP-1 metallo-beta-lactamase is limited to Acinetobacter, Alcaligenes, Citrobacter, Escherichia, Klebsiella, Pseudomonas, Salmonella, and Serratia.

Class 4 Integrons

Class 4 integrons are distinguished from other Resistance Integrons (RIs) by two key features; incorporation of hundreds of cassettes (V. cholerae, 216 unidentified genes in an array of 179 cassettes, holding roughly 3% of the genome) and the high similarity between the attC sites of those assembled cassettes. Class 4 integrons carry gene cassettes for antibiotics chloramphenicol and fosfomycin. Inspite of huge array of cassettes, identification of class 4 integron has been limited within members of Pseudomonas, Xanthomonas, Shewanella, Vibrionaceae, and other proteobacteria.

Integrative and Conjugative Elements

Integrative and conjugative elements (ICEs) were first of all anticipated by Burrus et.

al. are different mobile elements found in both Gram-positive and Gram-negative bacteria. These are self transmissible integrative elements that code for complete match of conjugation apparatus. ICEs can integrate into and excise from a host chromosome. These versatile entities support their own mobilization facilitating horizontal transfer of antibiotic-resistant genes, virulence factors and various bacterial traits. ICEs possess three genetic modules: (i) integration and excision module; (ii) conjugation module; and (iii) regulation module. These modules contain different array of genes that code for proteins operating by distinct mechanisms. ICEs contain a gene encoding an integrase (Int) that promotes site-specific integration and excision of the element, frequently into a unique site on the chromosome of the host organism. Some ICEs bear maintenance modules such as toxin–antitoxin systems and additional partition systems that guarantee thriving vertical inheritance of these elements. In contrast to plasmids, ICEs, are not found in extrachromosomal state, because they lack autonomous replication, the first known MGEs with ICE-like properties were Tn916 in Enterococcus faecalis and CTnDOT in Bacteriodes thetaiotaomicron). Bacteroides CTnDOT promote dissemination of antibiotic-resistant genes. ICEs are distinguished by element-specific properties although they possess a general life cycle and modular structure. Apart from resistance to antibiotics ICEs show a extensive collection of phenotypes on their hosts, including resistance for heavy metals and the power to degrade aromatic compounds. In addition, complex traits such as the capacity to inhabit a eukaryotic host fix nitrogen or encourage virulence and biofilm development have been recognized the connection between ICEs and the propagation of antibiotic resistance genes in some pathogens show that these mobile elements have considerable clinical significance. ICEberg is an integrated database that provides comprehensive information about integrative and conjugative elements (ICEs) found in bacteria.

Bacterial Toxin Anti-toxin Systems

Toxin-antitoxin (TA) systems, initially identified as plasmid addiction modules, are plentiful in the chromosomes of most free-living bacteria. TA systems provide endurance to bacterial populations in conditions of stress like nutrient deprivation or antibiotic pressure. Generally TA systems are made of a stable toxin and a labile antitoxin coded by a bicistronic locus. Toxin genes encode for proteins, while the antitoxin genes code for RNAs or antitoxin proteins, classifying them as type I–VI TA loci categorized due to mechanisms applied by the antitoxins to counteract the actions of the toxins. In TA systems I-VI product of the toxin gene is typically a protein, whereas the antitoxin gene is either a non-coding RNA among TA I and III or a low-molecular-weight protein in TA systems II, IV, V and VI. Toxins work on diverse targets to distress various cellular processes such as DNA replication, cell wall synthesis or protein synthesis. Amongst six types of TA system, type II is broadly studied, due to great quantity and high quality of freely accessible data. Currently two open-access bioinformatics resources in the field of type II TA loci, the online tool RASTA and the web-based database TADB are available.

Presently, a new toxin is reported that contains a Gcn5-related N-acetyltransferase (GNAT) domain that transfers the acetyl group from acetyl coenzyme A (Acetyl Co~A) to the amine group of tRNAs resulting in acetylation of tRNAs following inhibition of translation in bacterial cells. Similarly TacT of Salmonella enterica Typhimurium and AtaT of Escherichia coli O157:H7, also transfer the acetyl group from acetyl coenzyme A to the amine group of the tRNAs, In stress environment, the intensity of the alarmone molecule, (p)ppGpp, is amplified, which activates a particular proteinase that degrades the antitoxin by proteolytic cleavage, thus permitting the toxin to stop cell expansion. The free toxins then cause the dormant state of bacterial cells (persister cells) which can encourage bacterial tolerance to antibiotics. Persisters are a fraction of bacterial cells in the culture that survive through a prolonged antibiotic treatment. Many studies have shown that toxins of diverse chromosomal TA systems encourage the development of persister cells. GNAT-RHH TA system is a newly exposed approach of bacterial cells to support persister cell formation by disturbing tRNA functions, in S. enterica over expression of the TacT causes drug tolerance. Chromosomal type II TA loci have been reported to be activated by environmental stress. To assess if antibiotic stress would stimulate the transcription of kacAT operon, exponential phase HS11286-RR2 cells were checked with different antibiotics at the minimum inhibitory concentration (MIC). Transcript levels of the toxin gene, kacT, were quantified by RT-qPCR at various time points after antibiotic challenge. RT-qPCR results showed that the exposure to meropenem or tigecycline antibiotics caused 10-fold or 40-fold increase in kacT transcription levels.

Klebsiella pneumoniae faces a wide diversity of environmental conditions, including antibiotic stress. Fifteen pairs of putative type II TA loci are detected on the K. pneumoniae HS11286 chromosome. Activation of the toxin plays an important role in bacterial multidrug tolerance. The chromosomally encoded kacAT bicistronic operon of K. pneumoniae HS11286 is a functional GNAT-RHH TA locus with kacA encoding the cognate antitoxin to the toxic product of kacT.. Over expression of KacT inhibited K. pneumoniae cell growth and resulted in dormant cell formation. Crystal structure analyses show that the KacT toxin adapts a typical GNAT-fold, which may confer the same catalytic mechanism as the one revealed for TacT of S. enterica. It may bind cellular tRNAs via its positive groove and transfer the acetyl group from AcCoA to tRNAs halting translation leading to cell arrest.

Strategies to Combat Bacterial Antibiotic Resistance

Emerging antibiotic resistance is a problem of global magnitude. Confronted by increasing amounts of antibiotics use, emergence and dissemination of antibiotic resistant strains has compromised therapeutic potential of antibiotics. Of the different strategies adopted, techniques that materialize ideally include; (1) Designing antimicrobial peptides (AMPs; Bacteriocins, Cathelicidins, Microcins, etc.) with broader spectrum of targets, (2) Phage therapy (exploiting phages such as OMKO1, wksl3 and Φ1 to kill antibiotic-resistant bacteria), (3) combination therapy (using combination of antibiotics e.g., colistin in association with tigecycline, aminoglycoside, meropenem, etc or combination of inhibitor and antibiotic such as Augmention i.e., combination of clavulanate and amoxicillin), (4) Delivery of drugs, AMPs and essential oils as nanoparticles (NPs) for sustained and controlled release (AgNPs of penicillin G, amoxicillin, erythromycin, and vancomycin, (5) Liposomes as drug targeting vehicles, (6) Use of natural compounds such as Flavonoids, Alkaloids, Coumarins, etc, and (7) Modification of antimicrobials e.g., Plazomicin (ACHN-490); derivative of sisomicin produced by addition of a hydroxy-aminobutyric acid substituent at position 1 and a hydroxyethyl substituent at position 6. Other approaches include use of genomics to find out new bacterial targets and optimization of newer approaches that target bacterial pathogens while exerting selection for reduced pathogenesis, if bacteria evolve resistance to therapeutic intervention. Additionally, strategies such as designing molecules that can block bacterial attachment to surfaces and target bacterial virulence factors along with contribution to protect through production of inactivating antibodies, seems other suitable options to over the menace of drug resistance.

Bacterial infections continue to be one of the leading causes of morbidity and mortality worldwide. Fallout of excessive and imprudent antibiotic use, widespread dissemination of resistant determinants as part of MGEs has increased the rate of resistance development. Being capable to relocate between host genomes, they act as vehicles for resistance gene acquisition and their successive propagation. Thorough molecular studies have identified several mechanisms in microbes to attain the antimicrobial resistance. Among these mechanisms, plasmids, transposons, insertion sequences, integrons, ICEs and bacterial Toxin Anti-toxin systems have exposed how and why resistance has attained alarming stage. The possibility for recombination of genes from different bacterial populations is huge and it seems that it doesn't take bacteria much time to acquire the genetic resources to flourish in surroundings that would have otherwise hindered it's growth. Occurring with increasing frequency, resistance limits therapeutic option, resulting in the cases where certain human infections cannot be treated. Pertinently, where there is stiff resistance on the implementation of evidence-based clinical practice, scientists of the health care organizations are still searching as how to keep pace with the demand of actionable knowledge. This adverse condition of antimicrobial resistance demands the rejuvenation of dried pipeline for the development of new and efficient drugs to treat the deadly infection. With a goal to get hold of the menace of antibiotic resistance, it seems essential for everybody to have some basic knowledge about the systems in order to ensure optimal use of antibiotics from the surrounding milieu, to slow down the development of antibiotic-resistant superbugs.

ANTIBIOTIC RESISTANCE CRISIS

The rapid emergence of resistant bacteria is occurring worldwide, endangering the efficacy of antibiotics, which have transformed medicine and saved millions of lives.

Many decades after the first patients were treated with antibiotics, bacterial infections have again become a threat. The antibiotic resistance crisis has been attributed to the overuse and misuse of these medications, as well as a lack of new drug development by the pharmaceutical industry due to reduced economic incentives and challenging regulatory requirements. The Centers for Disease Control and Prevention (CDC) has classified a number of bacteria as presenting urgent, serious, and concerning threats, many of which are already responsible for placing a substantial clinical and financial burden on the U.S. health care system, patients, and their families. Coordinated efforts to implement new policies, renew research efforts, and pursue steps to manage the crisis are greatly needed.

Causes of the Antibiotic Resistance Crisis

Overuse

The overuse of antibiotics clearly drives the evolution of resistance. Epidemiological studies have demonstrated a direct relationship between antibiotic consumption and the emergence and dissemination of resistant bacteria strains.10 In bacteria, genes can be inherited from relatives or can be acquired from nonrelatives on mobile genetic elements such as plasmids. This horizontal gene transfer (HGT) can allow antibiotic resistance to be transferred among different species of bacteria. Resistance can also occur spontaneously through mutation. Antibiotics remove drug-sensitive competitors, leaving resistant bacteria behind to reproduce as a result of natural selection. Despite warnings regarding overuse, antibiotics are overprescribed worldwide.

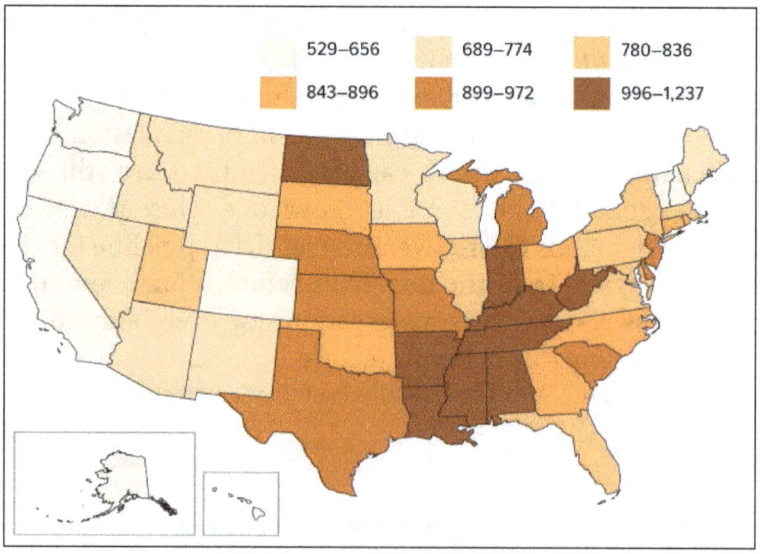

Antibiotic prescriptions per 1,000.

In the U.S., the sheer number of antibiotics prescribed indicates that a lot of work must be done to reduce the use of these medications. An analysis of the IMS Health Midas database, which estimates antibiotic consumption based on the volume of antibiotics sold in retail and hospital pharmacies, indicated that in 2010, 22.0 standard units (a unit equaling one dose, i.e., one pill, capsule, or ampoule) of antibiotics were prescribed per person in the U.S.17 The number of antibiotic prescriptions varies by state, with the most written in states running from the Great Lakes down to the Gulf Coast, whereas the West Coast has the lowest use. In some states, the number of prescribed courses of treatment with antibiotics per year exceed the population, amounting to more than one treatment per person per year.

The frequency with which doctors prescribe antibiotics varies greatly from state to state. The reasons for this variation are being studied and might suggest areas where improvements in antibiotic prescribing (fewer unnecessary prescriptions) would be most helpful.

In many other countries, antibiotics are unregulated and available over the counter without a prescription. This lack of regulation results in antibiotics that are easily accessible, plentiful, and cheap, which promotes overuse. The ability to purchase such products online has also made them accessible in countries where antibiotics are regulated.

Inappropriate Prescribing

Incorrectly prescribed antibiotics also contribute to the promotion of resistant bacteria.5 Studies have shown that treatment indication, choice of agent, or duration of antibiotic therapy is incorrect in 30% to 50% of cases. One U.S. study reported that a pathogen was defined in only 7.6% of 17,435 patients hospitalized with community-acquired pneumonia (CAP). In comparison, investigators at the Karolinska Institute in Sweden were able to identify the probable pathogen in 89% of patients with CAP through use of molecular diagnostic techniques (polymerase chain reaction [PCR] and semiquantitative PCR). In addition, 30% to 60% of the antibiotics prescribed in intensive care units (ICUs) have been found to be unnecessary, inappropriate, or suboptimal.

Incorrectly prescribed antibiotics have questionable therapeutic benefit and expose patients to potential complications of antibiotic therapy. Subinhibitory and subtherapeutic antibiotic concentrations can promote the development of antibiotic resistance by supporting genetic alterations, such as changes in gene expression, HGT, and mutagenesis. Changes in antibiotic-induced gene expression can increase virulence, while increased mutagenesis and HGT promote antibiotic resistance and spread. Low levels of antibiotics have been shown to contribute to strain diversification in organisms such as Pseudomonas aeruginosa. Subinhibitory concentrations of piperacillin and/or tazobactam have also been shown to induce broad proteomic alterations in Bacteroides fragilis.

Extensive Agricultural Use

In both the developed and developing world, antibiotics are widely used as growth supplements in livestock. An estimated 80% of antibiotics sold in the U.S. are used in animals, primarily to promote growth and to prevent infection. Treating livestock with antimicrobials is said to improve the overall health of the animals, producing larger yields and a higher-quality product.

The antibiotics used in livestock are ingested by humans when they consume food. The transfer of resistant bacteria to humans by farm animals was first noted more than 35 years ago, when high rates of antibiotic resistance were found in the intestinal flora of both farm animals and farmers. More recently, molecular detection methods have demonstrated that resistant bacteria in farm animals reach consumers through meat products. This occurs through the following sequence of events: 1) antibiotic use in food-producing animals kills or suppresses susceptible bacteria, allowing antibiotic-resistant bacteria to thrive; 2) resistant bacteria are transmitted to humans through the food supply; 3) these bacteria can cause infections in humans that may lead to adverse health consequences.

The agricultural use of antibiotics also affects the environmental microbiome. Up to 90% of the antibiotics given to livestock are excreted in urine and stool, then widely dispersed through fertilizer, groundwater, and surface runoff. In addition, tetracyclines and streptomycin are sprayed on fruit trees to act as pesticides in the western and southern U.S. While this application accounts for a much smaller proportion of overall antibiotic use, the resultant geographical spread can be considerable. This practice also contributes to the exposure of microorganisms in the environment to growth-inhibiting agents, altering the environmental ecology by increasing the proportion of resistant versus susceptible microorganisms.

Antibacterial products sold for hygienic or cleaning purposes may also contribute to this problem, since they may limit the development of immunities to environmental antigens in both children and adults. Consequently, immune-system versatility may be compromised, possibly increasing morbidity and mortality due to infections that wouldn't normally be virulent.

Availability of few New Antibiotics

The development of new antibiotics by the pharmaceutical industry, a strategy that had been effective at combating resistant bacteria in the past, had essentially stalled due to economic and regulatory obstacles. Of the 18 largest pharmaceutical companies, abandoned the antibiotic field. Mergers between pharmaceutical companies have also substantially reduced the number and diversity of research teams. Antibiotic research conducted in academia has been scaled back as a result of funding cuts due to the economic crisis.

Antibiotic development is no longer considered to be an economically wise investment for the pharmaceutical industry. Because antibiotics are used for relatively short periods and are often curative, antibiotics are not as profitable as drugs that treat chronic conditions, such as diabetes, psychiatric disorders, asthma, or gastroesophageal reflux.[1–3,13,14] A cost–benefit analysis by the Office of Health Economics in London calculated that the net present value (NPV) of a new antibiotic is only about $50 million, compared to approximately $1 billion for a drug used to treat a neuromuscular disease. Because medicines for chronic conditions are more profitable, pharmaceutical companies prefer to invest in them.

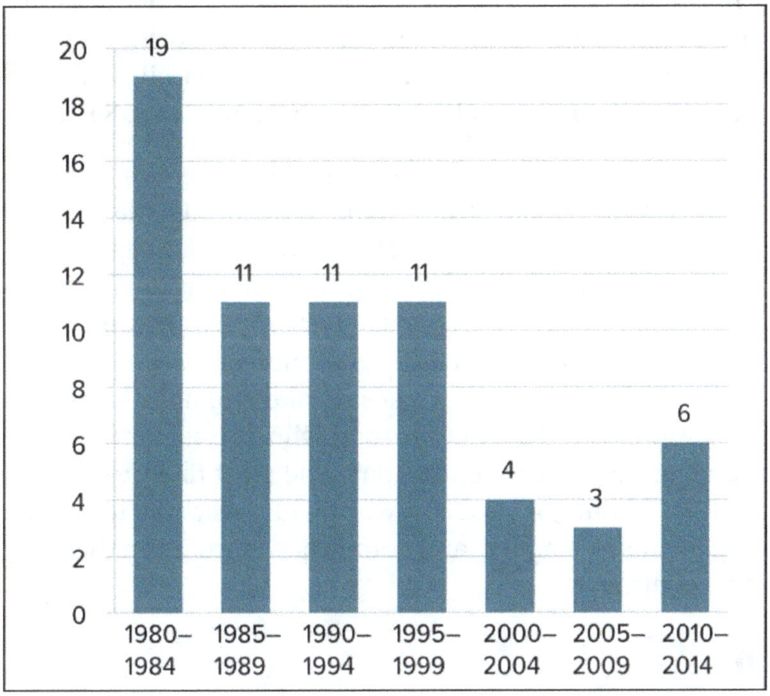

Number of antibacterial new drug application approvals versus year intervals.

Another factor that causes antibiotic development to lack economic appeal is the relatively low cost of antibiotics. Newer antibiotics are generally priced at a maximum of $1,000 to $3,000 per course compared with cancer chemotherapy that costs tens of thousands of dollars. The availability, ease of use, and generally low cost of antibiotics has also led to a perception of low value among payers and the public.

In addition, microbiologists and infectious-disease specialists have advised restraint regarding antibiotic use. Therefore, once a new antibiotic is marketed, physicians—rather than prescribing it immediately—often hold this new agent in reserve for only the worst cases due to fear of promoting drug resistance, and they continue to prescribe older agents that have shown comparable efficacy. Therefore, new antibiotics are often treated as "last-line" drugs to combat serious illnesses. This practice leads to the reduced use of new antibiotics and a diminished return on investment.

When new agents are eventually used, the emergence of resistance is nearly inevitable. However, since bacterial evolution is uncertain, the timeline for the development of resistance is unpredictable. A manufacturer that invests large sums of money into antibiotic development may therefore discover that profits are prematurely curtailed when resistance develops to a new antibiotic. Economic uncertainty related to the Great Recession has also had a restraining effect on the end users of antibiotics. Developed countries with well-funded health care systems have applied austerity measures, while developing countries such as China and India still have a large cohort of population that cannot afford expensive new medicines. As an additional complication, most antibiotics are currently off-patent and are supplied by manufacturers of generic drugs. The result has been access to cheap and generally effective drugs, which is good for the public; however, the downside is that many payers expect all antibiotics to be priced similarly—even new agents that target multidrug-resistant (MDR) pathogens.

Because of these factors, many large pharmaceutical companies fear a potential lack of return on the millions of U.S. dollars that would be required to develop a new antibiotic. The Infectious Diseases Society of America (IDSA) reported that as of 2013, few antibacterial compounds were in phase 2 or 3 development. In particular, the IDSA noted that unacceptably few agents with activity against emerging, extensively resistant gram-negative bacteria, such as Enterobacteriaceae, Pseudomonas aeruginosa, and Acinetobacter baumannii, were being developed. Pharmaceutical companies have also taken a more active interest in developing antibiotics for methicillin-resistant Staphylococcus aureus (MRSA), rather than gram-negative pathogens. The most likely explanation for this imbalance is that MRSA is a major problem worldwide, whereas the market for treating gram-negative organisms is smaller, and somewhat more unpredictable given that resistance is rapidly acquired.

Regulatory Barriers

Even for those companies that are optimistic about pursuing the discovery of new antibiotics, obtaining regulatory approval is often an obstacle. Between 1983 and 2007, a substantial reduction occurred in the number of new antibiotic approvals. Difficulties in pursuing regulatory approval that have been noted include: bureaucracy, absence of clarity, differences in clinical trial requirements among countries, changes in regulatory and licensing rules, and ineffective channels of communication.

Changes in standards for clinical trial design made by the U.S. Food and Drug Administration (FDA) during the past two decades have made antibiotic clinical trials particularly challenging. Studies comparing antibiotics with placebo are considered to be unethical; therefore, trials are designed to demonstrate noninferiority of new agents compared to existing drugs, within a varying statistical margin. This requires a large sample population and consequently high costs, making the development of antibiotics uneconomical and unattractive. While small companies have stepped in to fill the gap in antibiotic discovery and development formerly occupied by large pharmaceutical companies, the complexity and high cost of phase 3 clinical trials can exceed the financial means of these companies. However, in December 2014, Merck acquired the small antibiotic research company Cubist Pharmaceuticals, which is expected to accelerate the study and regulatory approval of new antibiotic agents in the future.

Altering the requirements for trial designs can have a significant impact on the size, and hence cost, of conducting clinical trials. Although more work in this area needs to be done, the FDA issued guidance in 2013 that changed the required clinical trial for acute bacterial skin and skin-structure infections. These changes included new disease state and endpoint definitions, a schedule for assessing endpoints, guidance on patient inclusion and exclusion, as well as supportive evidence and statistical justification for proposed noninferiority margins. Although still in draft form, the updated guidelines have been adopted in some clinical trials and serve as a basis for discussions regarding further study-protocol improvements.

Additional new regulatory approaches are needed to ensure the continued development and availability of antibiotic medications. The IDSA has proposed a new, limited-population antibiotic drug (LPAD) regulatory approval pathway that has drawn positive public comments from FDA officials. This model would enable substantially smaller, less-expensive, and faster clinical trials. In return for regulatory approval based on smaller clinical trials, the antibiotic would receive a very narrow indication focused only on the high-risk patients for whom benefits were shown to outweigh risks. Such limited approvals already exist in other situations, such as orphan drugs for the treatment of rare diseases.

Antibiotic-resistant Bacterial Infections

Antibiotic-resistant infections are already widespread in the U.S. and across the globe. Many public health organizations have described the rapid emergence of resistant bacteria as a "crisis" or "nightmare scenario" that could have "catastrophic consequences." The CDC declared in 2013 that the human race is now in the "post-antibiotic era," and in 2014, the World Health Organization (WHO) warned that the antibiotic resistance crisis is becoming dire. MDR bacteria have been declared a substantial threat to U.S. public health and national security by the IDSA and the Institute of Medicine, as well as the federal Interagency Task Force on Antimicrobial Resistance.

Among gram-positive pathogens, a global pandemic of resistant S. aureus and Enterococcus species currently poses the biggest threat. MRSA kills more Americans each year than HIV/AIDS, Parkinson's disease, emphysema, and homicide combined. Vancomycin-resistant enterococci (VRE) and a growing number of additional pathogens are developing resistance to many common antibiotics. The global spread of drug resistance among common respiratory pathogens, including Streptococcus pneumoniae and Mycobacterium tuberculosis, is epidemic.

Gram-negative pathogens are particularly worrisome because they are becoming resistant to nearly all the antibiotic drug options available, creating situations reminiscent of the pre-antibiotic era. The emergence of MDR (and increasingly pan-resistant) gram-negative bacilli has affected practice in every field of medicine. The most serious gram-negative infections occur in health care settings and are most commonly caused by Enterobacteriaceae (mostly Klebsiella pneumoniae), Pseudomonas aeruginosa, and

Acinetobacter. MDR gram-negative pathogens are also becoming increasingly prevalent in the community. These include extended-spectrum beta-lactamase-producing Escherichia coli and Neisseria gonorrhoeae.

The CDC assessed antibiotic-resistant bacterial infections according to seven factors: clinical impact, economic impact, incidence, 10-year projection of incidence, transmissibility, availability of effective antibiotics, and barriers to prevention. The threat level of each bacteria was then classified as "urgent," "serious," or "concerning". In general, threats that are urgent or serious require more monitoring and prevention activities, whereas those considered concerning require less.

CDC assessment of antibacterial resistance threats:

Urgent Threats

- Clostridium difficile.
- Carbapenem-resistant Enterobacteriaceae (CRE).
- Drug-resistant Neisseria gonorrhoeae.

Serious Threats

- Multidrug-resistant Acinetobacter.
- Drug-resistant Campylobacter.
- Fluconazole-resistant Candida (a fungus).
- Extended spectrum beta-lactamase-producing Enterobacteriaceae (ESBLs).
- Vancomycin-resistant Enterococci (VRE).
- Multidrug-resistant Pseudomonas aeruginosa.
- Drug-resistant nontyphoidal Salmonella.
- Drug-resistant Salmonella Typhimurium.
- Drug-resistant Shigella.
- Methicillin-resistant Staphylococcus aureus (MRSA).
- Drug-resistant Streptococcus pneumoniae.
- Drug-resistant tuberculosis.

Concerning Threats

- Vancomycin-resistant Staphylococcus aureus (VRSA).

- Erythromycin-resistant Group A Streptococcus.
- Clindamycin-resistant Group B Streptococcus.

Methicillin-resistant Staphylococcus Aureus

MRSA was first identified five decades ago. Since then, MRSA infections have spread worldwide, appearing at a high incidence in several countries in Europe, the Americas, and the Asia-Pacific region. MRSA infections can be very serious and are among the most frequently occurring of all antibiotic-resistant threats. In the U.S., 11,285 deaths per year have been attributed to MRSA alone.

MRSA is resistant to penicillin-like beta-lactam antibiotics. However, a number of drugs still retain activity against MRSA, including glycopeptides (e.g., vancomycin and teicoplanin), linezolid, tigecycline, daptomycin, and even some new beta-lactams, such as ceftaroline and ceftobiprole. However, MRSA has shown outstanding versatility at emerging and spreading in different epidemiological settings over time (in hospitals, the community, and, more recently, in animals). This compounds the epidemiology of MRSA infections and creates a challenge for infection-control systems that focus only on health care–associated infections (HAIs). Moreover, although resistance to anti-MRSA agents usually occurs through bacterial mutation, there have been reports of the transfer of resistance to linezolid and glycopeptide antibiotics, which is cause for major concern.

Fortunately, the incidence of HAI MRSA infections seems to be declining, since aggressive preventive hygiene measures in hospitals in some areas (i.e., the Netherlands and United Kingdom) have had a positive effect. Between 2005 and 2011, overall rates of invasive MRSA dropped 31%; the largest declines (around 54%) were observed in HAIs. This outcome provides evidence that infection control can be highly effective at limiting the spread of MRSA. However, during the past decade, rates of community-acquired MRSA infections have increased rapidly among the general population. While there is some evidence that these increases are slowing, they are not following the same downward trends that have been observed for hospital-acquired MRSA infections.

Vancomycin-resistant Enterococci

VRE presents a major therapeutic challenge. Enterococci cause a wide range of illnesses, mostly among patients in hospitals or other health care settings, including bloodstream, surgical-site, and urinary tract infections. VRE infections, often caused by Enterococcus faecium and less frequently by Enterococcus faecalis, have a lower prevalence and epidemiological impact than MRSA does worldwide, except for the U.S. and some European countries. An estimated 66,000 HAI Enterococci infections occur in the U.S. each year. The proportion of infections that are vancomycin-resistant depends on the species. Overall, 20,000 (30%) of hospital-acquired enterococcal infections per year are vancomycin-resistant, leading to 1,300 deaths.

Few antimicrobial options are available to treat VRE.16 Antibiotics used against VRE include linezolid and quinupristin/dalfopristin, while the role of daptomycin and tigecycline needs to be further defined. VRE remains a major threat; consequently there is tremendous interest in developing novel drugs that could have bactericidal activity against VRE, such as oritavancin.

Drug-resistant Streptococcus Pneumoniae

S. pneumoniae can cause serious and sometimes life- threatening infections. It is a major cause of bacterial pneumonia and meningitis, as well as bloodstream, ear, and sinus infections. Resistant S. pneumoniae infections complicate medical treatment, resulting in nearly 1.2 million illnesses and 7,000 deaths per year. The majority of these cases and deaths occur among adults 50 years of age or older, with the highest rates among those 65 years of age or older. S. pneumoniae has developed resistance to drugs in the penicillin class and erythromycins, such as amoxicillin and azithromycin, respectively. It has also developed resistance to less commonly used drugs. In 30% of severe S. pneumoniae cases, the bacteria are fully resistant to one or more clinically relevant antibiotics.

Fortunately, a new version of pneumococcal conjugate vaccine (PCV13), introduced in 2010, protects against infections caused by the most resistant pneumococcus strains, so rates of resistant S. pneumoniae infections are declining. From 2000 to 2009, an earlier pneumococcal conjugate vaccine, PCV7, provided protection against seven pneumococcal strains, but PCV13 expanded this protection to 13 strains. Use of this vaccine has not only prevented pneumococcal disease, it has also reduced antibiotic resistance by blocking the transmission of resistant S. pneumoniae strains.

Drug-resistant Mycobacterium Tuberculosis

Drug-resistant M. tuberculosis infections are a serious threat in the U.S., and an even more urgent threat worldwide. The WHO reported that in 2012, 170,000 people died from drug-resistant tuberculosis (TB) infections. M. tuberculosis is most commonly spread through the air. Infections caused by this bacterium can occur anywhere in the body but most often appear in the lungs. Of a total of 10,528 TB cases reported in the U.S. in 2011, antibiotic resistance was identified in 1,042, or 9.9%. The major factors driving TB drug resistance are incomplete, incorrect, or unavailable treatment and a lack of new drugs.

In most instances, TB infections are treatable and curable with available first-line drugs, such as isoniazid or rifampicin; however, in some cases, M. tuberculosis can be resistant to one or more of these first-line drugs. Treatment of drug-resistant TB can be complex, requiring longer treatment periods and more expensive drugs that often have more side effects. Extensively drug-resistant TB (XDR-TB) is resistant to most TB drugs, including isoniazid and rifampicin, any fluoroquinolones, and any of the three second-line injectable drugs (i.e., amikacin, kanamycin, and capreomycin);

therefore, fewer treatment options are available for patients with XDR-TB, and the drugs that are available are much less effective. Although drug-resistant TB and XDR-TB infections are an increasing threat worldwide, these infections are uncommon in the U.S. because of the implementation of a robust TB infection prevention and management program.

Carbapenem-resistant Enterobacteriaceae

Carbapenem-resistant Enterobacteriaceae (CRE) are a group of bacteria that have become resistant to "all or nearly all" available antibiotics, including carbapenems, which are typically reserved as the "treatment of last resort" against drug-resistant pathogens. An enzyme called New Delhi metallo-beta-lactamase (NDM-1) is present in some gram-negative Enterobacteriaceae bacteria (notably Escherichia coli and K. pneumoniae) that makes them resistant to virtually all beta-lactams, including carbapenems.

Untreatable or difficult-to-treat infections due to CRE bacteria are on the rise among patients in medical facilities. An estimated 140,000 health care–associated Enterobacteriaceae infections occur in the U.S. each year; 9,300 of these are caused by CRE.5 Each year, approximately 600 deaths result from infections caused by the two most common types of CRE, carbapenem-resistant Klebsiella species and carbapenem-resistant E. coli.

MDR Pseudomonas Aeruginosa

P. aeruginosa is a common cause of HAIs, including pneumonia and bloodstream, urinary tract, and surgical-site infections. More than 6,000 (13%) of the 51,000 health care–associated P. aeruginosa infections that occur in the U.S. each year are MDR. Roughly 400 deaths per year are attributed to these infections. Some strains of MDR P. aeruginosa have been found to be resistant to nearly all antibiotics, including aminoglycosides, cephalosporins, fluoroquinolones, and carbapenems.

MDR Acinetobacter

Acinetobacter is a gram-negative bacterium that causes pneumonia or bloodstream infections, especially in critically ill patients on mechanical ventilation. Some Acinetobacter species have become resistant to all or nearly all antibiotics, including carbapenems, which are often considered to be the drug of last resort. About 12,000 health care–acquired Acinetobacter infections occur in the U.S. each year, and 7,300 (63%) of these are MDR (resistant to at least three different classes of antibiotics), causing 500 deaths per year.

ESBL-producing Enterobacteriaceae

Extended-spectrum beta-lactamase (ESBL)-producing Enterobacteriaceae carry a broad-spectrum beta-lactamase enzyme that enables them to become resistant to a wide variety of penicillin and cephalosporin antibiotics. ESBL-producing Enterobacteriaceae cause 26,000 HAIs and ,700 deaths per year. Some ESBL-producing Enterobacteriaceae are resistant to nearly all antibiotics in the penicillin and cephalosporin classes. In such cases, the remaining treatment option is an antibiotic from the carbapenem family. However, these drugs should be used with caution, since use contributes to resistance.

Drug-resistant Neisseria Gonorrhoeae

In recent years, drug-resistant forms of N. gonorrhoeae, the causative agent for the sexually transmitted disease gonorrhea, have begun to emerge in the U.S. Gonorrhea is characterized by discharge and inflammation of the urethra, cervix, pharynx, or rectum. While not normally fatal, gonorrhea spreads easily and can cause severe complications in reproductive functions. The CDC estimates that more than 800,000 cases of gonorrhea occur annually, making it the second-most-frequently reported infectious disease in the U.S.5 Should drug-resistant N. gonorrhoeae become more widespread, it has been estimated that it would cause 75,000 additional cases of pelvic inflammatory disease, 15,000 cases of epididymitis, and 222 additional human immunodeficiency virus infections over a projected 10-year period.

Cephalosporin-resistant N. gonorrhoeae is often resistant to other types of antibiotics, such as fluoroquinolones, tetracyclines, and penicillins. Infections caused by these bacteria will therefore likely fail empiric treatment regimens. In response to this challenge, the CDC has updated its treatment guidelines to recommend ceftriaxone, plus either azithromycin or doxycycline, as the first-line treatment for gonorrhea.

Clinical and Economic Burden of Antibiotic Resistance

Antibiotic-resistant infections are a substantial health and economic burden to the U.S. health care system, as well as to patients and their families. They commonly occur in hospitals, due to the clustering of highly vulnerable patients, extensive use of invasive procedures, and high rates of antibiotic use in this setting. Nearly two million Americans per year develop HAIs, resulting in 99,000 deaths, most due to antibacterial-resistant pathogens. In 2006, two common HAIs (sepsis and pneumonia) were found to be responsible for the deaths of nearly 50,000 Americans and cost the U.S. health care system more than $8 billion.

Antibiotic-resistant infections add considerable costs to the nation's already overburdened health care system. When first-line and then second-line antibiotic treatment options are limited or unavailable, health care professionals may be forced to use antibiotics that are more toxic to the patient and frequently more expensive. Even when effective treatments exist, data show that in most cases patients with resistant infections require significantly longer hospital stays, more doctors visits, and lengthier recuperations and experience a higher incidence of long-term disability. The duration of hospital stays for patients with antibiotic-resistant infections was found to be prolonged by 6.4 to 12.7 days, collectively adding an extra eight million hospital days.

Estimates regarding the medical cost per patient with an antibiotic-resistant infection range from $18,588 to $29,069. The total economic burden placed on the U.S. economy by antibiotic-resistant infections has been estimated to be as high as $20 billion in health care costs and $35 billion a year in lost productivity. Antibiotic-resistant infections also burden families and communities due to lost wages and health care costs.

2

Types of Antimicrobial Drug

Antimicrobial drugs are used for treating a microbial infection. There are various types of antimicrobial drugs which can be named as antiviral drug, antifungal medication, antiparasitics and antibiotic medication. The topics elaborated in this chapter will help in gaining a better perspective about these types of antimicrobial drugs.

Antibiotics play a vital role in dental practice for managing orofacial infections. They are used to manage existing infection and they are also used as prophylaxis for certain medical conditions and surgical procedures.

An astounding number of drug classes and formulations are available to manage infections. Fortunately, the microorganisms associated with odontogenic and periodontal infections are well characterized and a relatively small number of antimicrobial agents are required to effectively manage dental infections. With the exception of allergy, adverse effects attributed to these antibiotics are surprisingly infrequent, but most agents are implicated in producing nausea, dyspepsia, and diarrhea.

Pharmacological Considerations

Antibiotics are generally classified according to their molecular structure and their antimicrobial mechanisms. Ideally, these mechanisms of action either interrupt synthesis of structural components or alter specific metabolic functions that are unique to microbial cells. Such specificity cannot always be accomplished, but if it is, human cells can be spared cytotoxic effects.

Beta Lactam Derivatives

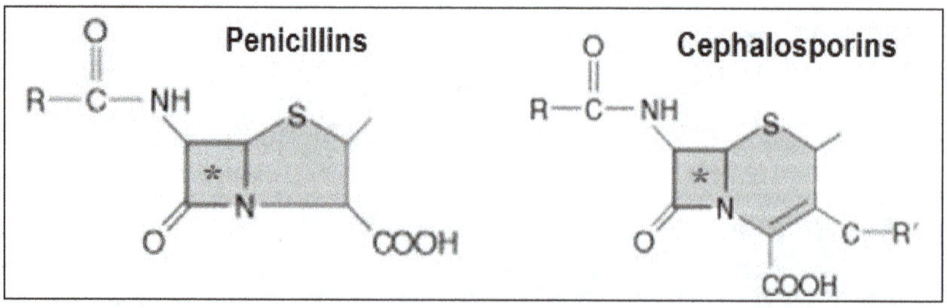

Beta lactam structure. Penicillins and cephalosporins contain a beta lactam ring (asterisk) that conveys their antimicrobial action and is the target for microbial resistance.

The molecular structures of penicillins and cephalosporins have a beta lactam ring in common that accounts for their ability to inhibit cell wall synthesis in susceptible microorganisms. Additional beta lactam derivatives include the monobactams and carbapenems, but these are not used for dental infections. The beta lactam ring is also the target for resistant species, which produce a variety of beta lactamases formerly called penicillinase and/or cephalosporinase. Prominent microbial species that produce beta lactamases are Staphylococcus aureus and Haemophilus influenzae. However, several species of Bacteroides and Prevotella, which contribute to the more severe odontogenic and periodontal infections, may also demonstrate this mechanism of resistance.

The similar molecular structure for penicillins and cephalosporins raises concern regarding cross-allergenicity. Formerly this was thought to be related to the beta lactam ring, but recent evidence has established that cross-allergenicity is related more to similarities in the R side chains. It is generally accepted that patients having a history of IgE-mediated reaction to a penicillin should be managed using a non–beta lactam antibiotic. Urticaria (hives) is immunoglobulin E mediated but accounts for only 10% of all exanthematous drug reactions. The overwhelming majority of cutaneous reactions to penicillins are pruritus or rash; these are not immunoglobulin E mediated, and any potential for cross-reaction is unlikely. Furthermore, upon further questioning, most patients claiming allergy to penicillin are discovered to have experienced stomach upset (dyspepsia), nausea, or diarrhea. Although macrolides and clindamycin are conventionally considered the alternatives of choice in patients allergic to penicillins, the macrolides have become less attractive. It is preferable to substitute an alternate penicillin or cephalosporin for a penicillin-allergic patient, provided the nature of the patient's reaction was merely pruritic (itch) or a maculopapular rash. A history of urticaria (hives) or anaphylactoid symptoms is more convincing evidence that the patient's reaction to penicillin was truly immunoglobulin E mediated, and in this case, one should refrain from prescribing any beta lactam derivative. Clinical management of the penicillin-allergic patient is summarized in figure.

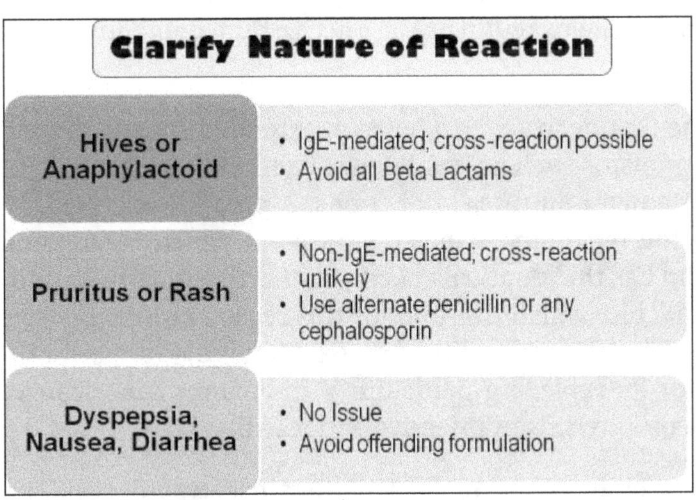

Managing penicillin-allergic patients.

Penicillins

The fortuitous discovery of penicillin is credited to Alexander Fleming, and its clinical use began in 1941. This first penicillin was designated penicillin G and its dosage expressed in units or milligrams, 1600 units approximating 1 mg. It is very active against gram-positive cocci that frequently cause oral, pharyngeal, and pulmonary infections. It is also active against Neisseria gonorrhoeae and Treponema pallidum. For this reason, penicillin G is still a first-line agent for treating the 2 most familiar venereal diseases, syphilis and gonorrhea. Penicillin G is highly degradable in gastric acid and is generally administered only by parenteral routes, formulated as several salts that differ in their rate of absorption. Potassium and sodium salts are used for intravenous administration and procaine or benzathine salts for intramuscular injection. Benzathine formulations provide low serum levels for 3–4 weeks and are generally reserved for situations in which patient compliance with oral administration is unlikely.

The phenoxymethyl derivative of penicillin, designated penicillin V, is more acid stable and has become the standard penicillin for oral use. Its spectrum is similar to that for penicillin G, but it is less active against Neisseria species and several anaerobes. It remains the agent of choice for mild to moderate intraoral infections but is susceptible to a variety of beta lactamases produced by most strains of S aureus and many species of Bacteroides. Although the former are seldom causative agents in oral infections, resistant strains of Bacteroides represent a significant cause for penicillin failure. This includes several species of Bacteroides that have been reclassified as other genera, either Porphyromonas or Prevotella.

To counter microbial resistance attributed to beta lactamase, derivatives of penicillin have been synthesized that are not susceptible to enzymatic breakdown. These penicillinase-stable agents include oxacillin and dicloxacillin. They are primarily indicated in the management of infections attributed to S aureus, over 90% of which are resistant to penicillin G and V. These microbes are rarely if ever present in odontogenic infections, and both oxacillin and dicloxacillin are less active than penicillin V against odontogenic pathogens.

Ampicillin was the first derivative to have an extended spectrum that includes several gram-negative organisms such as H influenzae and Escherichia coli, but these are rarely, if ever, associated with intraoral infections. A possible exception may be infections that also involve the maxillary or nasal sinuses in which H influenzae is a common culprit. Amoxicillin has the identical spectrum of activity attributed to ampicillin, but it exhibits greater oral bioavailability, and for this reason ampicillin is generally reserved for parenteral use. The bioavailability of amoxicillin is also superior to that of penicillin V and accounts for its replacing penicillin V in current American Heart Association guidelines for the prophylaxis of infective endocarditis.

Amoxicillin and penicillin V are equally active against streptococci. The only advantages of amoxicillin for dental infections are greater bioavailability and a longer half-life. For this reason, if one adheres to the principle of choosing antibiotics having the narrowest spectrum possible, penicillin V is preferred over amoxicillin. Routine use of amoxicillin fosters the accumulation of amoxicillin-resistant microorganisms that lack dental implications. However, if one considers that the leading cause of therapeutic failure is lack of patient compliance, superior bioavailability and less frequent dosing (3 times daily [TID] versus 4 times daily [QID] for penicillin V) favor the use of amoxicillin.

Ampicillin and amoxicillin are not beta lactamase stable and are available in formulations that include so-called beta lactamase inhibitors, sulbactam and clavulanic acid respectively. Clavulanic acid has a molecular structure similar to that of penicillin and has weak antimicrobial action but strong affinity for beta lactamase. It is combined with amoxicillin to act as a "suicide molecule" protecting it from beta lactamase. The combination of amoxicillin and clavulanic acid (Augmentin) is very expensive and should not be used for routine empiric therapy. Resistance among streptococci and most other dental pathogens is not attributed to beta lactamase. At best, the use of this product may be a heroic option in refractory odontogenic and periodontal infections that, on some occasions, become colonized by penicillin-resistant Bacteroides and Prevotella species that produce this enzyme.

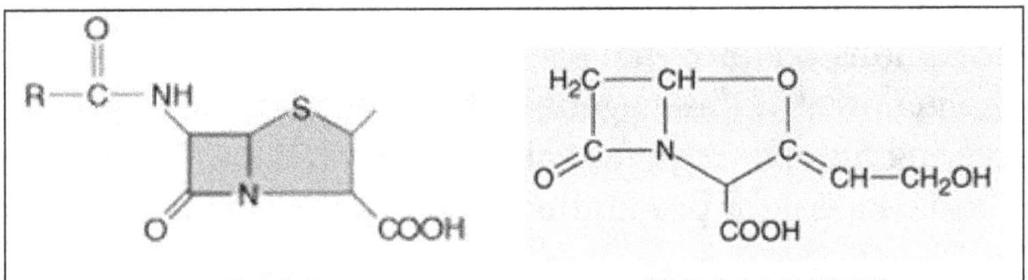

Clavulanic acid. The molecular structure of clavulanic acid resembles that of penicillin enough to compete for beta lactamases that are secreted by resistant microorganisms. This limits the amount of enzyme available to inactivate the penicillin.

Cephalosporins

The first generation of cephalosporins has a spectrum of activity that includes that of penicillin V for odontogenic microbes. They are also active against most strains of S aureus because, unlike penicillin, they are not susceptible to beta lactamases produced by this species. S aureus is a common nosocomial pathogen, and for this reason the cephalosporins are ordered more frequently than the penicillins for hospitalized patients. First-generation cephalosporins offer few advantages over penicillins in the management of dental infections. they are an alternative for the penicillin-allergic patient, and certain agents have pharmacokinetic advantages that allow less frequent dosing. For example, the long elimination half-life of cefadroxil (Duricef) allows twice daily (BID) oral dosing, whereas cephalexin (Keflex) is administered QID. Cefazolin (Ancef, Kefzol) is the standard first-generation agent for parenteral use.

Second- and third-generation cephalosporins exhibit an even broader spectrum and greater resistance to beta lactamase. Several of the third-generation agents also demonstrate antipseudomonal activity. Although these antibiotics have valuable indications in the management of specific infectious diseases, too often they are prescribed inappropriately for infections that could be managed using less expensive agents. They are rarely if ever indicated for managing oral infections.

Macrolides

Macrolide antibiotics exert their antibacterial action by binding to the 50S ribosomal subunit of susceptible organisms, resulting in inhibition of protein synthesis. Erythromycin is the prototypic macrolide and has been prescribed historically as an alternative for patients allergic to penicillin because it formerly had reasonable activity against most penicillin-sensitive microbes. This is no longer the case, however. Macrolides have little activity against periodontal pathogens, and in recent years their activity against streptococcal species has declined to such a degree that most experts discourage their use in odontogenic infections as well. It should be noted that streptococci and staphylococci that are resistant to erythromycin are also resistant to clarithromycin and azithromycin. Clarithromycin and azithromycin are more active than erythromycin against Streptococcus viridans, which accounts for their mention as penicillin alternatives in current American Heart Association guidelines for prophylaxis of infective endocarditis. In addition to growing resistance among odontogenic pathogens, macrolides produce a high incidence of nausea, and the majority of these agents inhibit cytochrome P450 enzymes. This accounts for a staggering number of potential drug interactions. Furthermore, azithromycin continues to be a concern for promoting cardiac arrhythmias in susceptible patients.

Tetracyclines

Like the macrolides, tetracyclines are bacteriostatic but exert their antimicrobial effect by binding to the 30S subunit of the bacterial ribosome to inhibit protein synthesis. Tetracyclines have a wide spectrum of activity, but microbial resistance has increased to the extent that they are seldom first-line agents for medically treated infections. Sinus and respiratory infections caused by H influenzae and pneumonococci are notable exceptions because most of these strains remain sensitive.

In dental practice, tetracyclines are useful adjuncts for managing periodontal infections. They are highly active against many of the microorganisms implicated in gingival and periodontal disease and exhibit high bioavailability in the gingival sulcus. However, their efficacy is unreliable for managing odontogenic infections due to streptococcal resistance. Notably, there is less antimicrobial resistance to doxycycline and minocycline than to other, more conventional tetracyclines, and these are generally preferred for periodontal infection.

Like other tetracyclines, doxycycline (Vibramycin) increases skin sensitivity to sunlight, leading to intense sunburn and generalized erythema, but offers several advantages. It is well absorbed in the presence of food and has an extended elimination half-life that allows for once-daily dosing. It is eliminated primarily in feces, and this makes it particularly attractive for patients having hepatic or renal compromise. Finally, even at subantimicrobial doses, doxycycline inhibits the activity of collagenases that contribute to the pathogenesis of periodontal destruction.

Metronidazole

Metronidazole is a prodrug that is converted to a toxic radical within anaerobic microbes. The radical destroys existing DNA and other vital compounds, rendering it bactericidal against most anaerobic organisms. For this reason, it is very useful for treating severe odontogenic and periodontal infections where anaerobes are able to thrive. It is not recommended as monotherapy for oral infections because it is inactive against aerobic and facultative streptococci. However, it may be combined with beta lactams when managing severe refractory infections. Patients should be cautioned to avoid alcoholic beverages while taking this medication because disulfiram-like reactions have been reported. These consist of severe nausea and abdominal cramping due to the formation of a toxic compound resembling formaldehyde.

Clindamycin

Clindamycin binds to the 50S subunit of bacterial ribosomes to suppress protein synthesis, but, unlike the macrolides, it is bacteriocidal.1 It has reliable activity against both aerobic and anaerobic cocci, as well as most species of Bacteroides, including Bacteroides fragilis. These pathogens are often implicated in severe orofacial infections. Its cost and predilection for Clostridium difficile infection limit its routine use for dental infections in favor of beta lactams. However, these should not be deterrents to using clindamycin when indicated. Indeed, C difficile infection may be a complication associated with amoxicillin and cephalosporins as well.

Antifungal Agents

Nystatin was formerly the most common antifungal agent used for oral candida infections but today the azole derivatives are preferred. These agents inhibit the synthesis of ergosterol, an essential component of the fungal cell membrane. Although a variety of agents are available, clotrimazole (Mycelex) troches are generally preferred based on cost and little risk for side effects and drug interactions. Clotrimazole can be prescribed as 10-mg troches administered 5 times daily for 10–14 days. Fluconazole (Diflucan) is available for oral (PO) administration and miconazole (Oravig) as once-daily buccal tablets, but they are very expensive. These 2 azoles also inhibit several families of cytochrome P450 enzymes and should be avoided in patients taking warfarin, statins, antiretrovirals, and any drug known to prolong QT intervals.

Miscellaneous Agents

The following agents have little or no indications for managing odontogenic or periodontal infections. However, some familiarity is essential for collaborative medical management of dental patients. The most significant risk for infective endocarditis is in patients having a prior history of this infection, especially those with prosthetic valves. In these cases, some cardiologists may prefer that an aminoglycoside such as gentamicin be added to the prophylactic regimen. This is based on the synergistic influence these agents have with cell wall inhibitors in killing enterococci and resistant strains of S viridans. However, there is little scientific evidence to support this practice. For high-risk patients who are allergic to beta lactams, vancomycin may be requested and does not require the addition of gentamicin. These antibiotics must be administered by intravenous infusion over 30–60 minutes preoperatively, which may require that treatment be arranged at an outpatient clinic if the provider is inexperienced with this route of administration.

Aminoglycosides

Historically, the most familiar member of this class is streptomycin, formerly the agent of choice for treating tuberculosis. Gentamicin and tobramycin are used most commonly and are the primary agents used to treat infections caused by gram-negative rods, most notably Pseudomonas species. Although most antibiotics that inhibit protein synthesis are bacteriostatic, the aminoglycosides are frequently bactericidal. The newer generations of beta lactam antibiotics are also active against Pseudomonas species, but they are far more expensive.

The major drawback of the aminoglycosides is their toxicity. This includes nephrotoxicity and ototoxicity. Nephrotoxicity is fairly common, but is generally reversible following discontinuation of the offending drug. Ototoxicity may include either the auditory or vestibular portions of the eighth cranial nerve. Unlike renal cells, sensory neurons cannot regenerate, and toxicity may be permanent. Hearing and labyrinthine function must be carefully monitored because initial symptoms may be reversible provided the drug is withdrawn quickly. Tinnitus is the earliest sign of auditory toxicity, whereas headache and nausea generally herald the onset of vestibular toxicity. The various preparations differ in their propensity to induce toxicity in each division.

Vancomycin

Vancomycin inhibits cell wall synthesis and is active against most gram-positive cocci, including most species of streptococci, staphylococci, and enterococci.1 Although enterococci were once uniformly susceptible to vancomycin, outbreaks of infections caused by resistant strains are a growing problem. Scattered cases of infections caused by vancomycin-resistant enterococci are particularly sobering because they frequently result in mortality. Consequently, vancomycin is reserved for treatment of serious infections caused by organisms that are resistant to first-line agents, such as beta lactams, or in cases of serious allergy to these antibiotics. This may include patients at great risk for infective endocarditis who are allergic to beta lactams, as an alternative to clindamycin.

Vancomycin can produce pseudoallergic reactions. Rapid intravenous infusion may trigger histamine release, which causes a variety of symptoms, including erythematous or urticarial reactions, flushing, tachycardia, and hypotension. The extreme flushing is called red-man syndrome. The most significant untoward reactions are ototoxicity and nephrotoxicity. Auditory impairment can be permanent, especially with excessively high plasma concentrations. Nephrotoxicity is less common, but renal function should be monitored and appropriate plasma concentrations maintained. As might be expected, toxicity is more likely when administered concurrently with an aminoglycoside.

Fluoroquinolones

The fluoroquinolones are synthetic, broad-spectrum antibacterial agents that inhibit DNA gyrase, an essential enzyme that is involved in the replication, transcription, and repair of bacterial DNA. The introduction of quinolone derivatives is the most significant recent advance in antimicrobial therapy. Ciprofloxacin (Cipro) was the first of these agents introduced and is regarded as the prototype. It is active against most staphylococci and a variety of gram-negative microorganisms, but has poor activity against most streptococci and all anaerobes. This negates its use for odontogenic and periodontal infections, which generally consist of mixed aerobic and anaerobic flora. Newer generations of quinolones have broader activity, but their cost generally renders them inappropriate for dental-related infections. Levofloxacin (Levaquin) has good antistreptococci action but poor anaerobic activity. Once daily dosing is attractive, but its cost is ~$25/d and its extensive antimicrobial spectrum completely overrules conventional principles of antibiotic prescribing for odontogenic infections. Gemifloxacin (Factive) offers added anaerobic coverage but at a cost of ~$50/d. At best it might be argued as an outrageously expensive effort to heroically salvage a persistent peri-implant infection.

Table: Antibiotic susceptibility data for selected antibiotics.

Data	Penicilli V	Amoxiciflin	Cephalexin	Clindomycin	CiproJloxacin
PO dose (m9)	500	500	500	300	500
Peak serum level (pg/mL)	4	7.5	5.8	3.6	2.4
$T_{1/2}$ (h)	0.5 – 0.75	1.4-2.0	0.5-1.2	2.4	4
MIC-90 (μg/mL)					
Streptococcus viridans	0.01	0.05	0.12	0.06-256	4
Peptostreptococcus	0.5	0.2	†	0.1→J00	2-6.25
Prevotella	0.5	0.5	‡	0.03→128	16→64
Haemophilus influenzae ‡	0.8	0.5	16	0.4-50	0.015-0.06
Bacteroides frogilis§	>32	>32	>32	0.03→ 128	4→64

"MIC indicates minimum inhibitory concentration; PO, oral: $T_{1/2}$, elimination half-life. The MIC-90 of penicillin V for S viridans is 0.01 μg/mL. Its peak serum concentration following oral administration of 500 mg is 4 μg/mL and its serum half-life is 0.5-0.75 hours. Following its administration and allowing 1 hour for peak concentration to be achieved, serum concentrations of penicillin V will remain above the MIC-90 for ~6—7 half-lives. For this reason, penicillin V is considered highly effective for managing most odontogenic infections. Notice that this would not be the case for managing severe dental infections attributed to B fragilis, but clindamycin is effective. Also note the basis for amoxicillin being prescribed 3 times daily rather than 4 times daily for penicillin V and moreover that fluoroquinolones such as ciprofloxacin are ineffective for dental pathogens.

Antibiotic Complications

There are surprisingly few complications associated with antibiotic therapy. Other than allergy to beta lactam agents addressed previously, the principal complications are gastrointestinal related and potential drug interactions.

Antibiotic-associated Diarrhea and Colitis

Antibiotic-associated diarrhea is not that uncommon during a course of antibiotic therapy, but it becomes a more significant event if it is the result of C difficile infection, a common nosocomial pathogen. This anaerobic, spore-forming bacillus is most often contracted during hospitalization or prolonged stays in other health care facilities. However, it is also found in the community, and this source of infection is on the rise. Intestinal flora normally prevent colonization by C difficile, and it is present in only 1–4% of the general population, but >20% in those admitted to health care facilities for a week or more. When normal flora is altered by antibiotic therapy and the patient either harbors or comes into contact with C difficile, colonization increases. Colonization may be enhanced by most antibiotics, but clindamycin, amoxicillin, second- and third-generation cephalosporins, and the fluoroquinolones are most often implicated.

Colonization alone does not necessarily result in C difficile infection. Risk for actual infection depends on the interaction of several additional factors, including virulence of the particular strain and patient-related factors such as age, immune status, and the concurrent use of acid-reduction gastrointestinal drugs, eg, proton pump inhibitors. Once C difficile infection occurs, the consequences range from diarrhea to pseudomembranous colitis. Therefore, in outpatient dental practice the typical sequence of events leading to C difficile infection are as follows:

- The patient is currently colonized with C difficile. (This is most likely if the patient has recently visited, has been a patient, or is a health care provider in a hospital or nursing home.)

- Colonization is then increased by an antibiotic altering intestinal flora. (Clindamycin or amoxicillin are most likely.)

- Patient-related factors determine risk for actual infection and subsequent severity. (Many variables contribute, but older age, poor immune status, and use of acid-reduction drugs are most significant.)

The incidence of diarrhea attributed to those antibiotics commonly used in dentistry ranges from 2 to 10% and may be as high as 25% with amoxicillin/clavulanic acid (Augmentin). Clinically, the challenge is to distinguish this so-called nuisance diarrhea from that associated with C difficile disease. In a patient who normally tolerates antibiotics but experiences diarrhea that is florid (≥3 unformed stools per day for ≥2 days) and complains of abdominal pain, C difficile infection should be suspected. Milder symptoms in patients who have previously experienced diarrhea with antibiotics favor nuisance diarrhea.

Mild, nuisance diarrhea may be managed using antiperistaltics and changing the antibiotic to a narrower spectrum if possible. However, if diarrhea is severe and C difficile infection is suspected, the following is suggested:

- Avoid antiperistaltics. Accumulation of toxin can worsen the infection.

- Stop the current antibiotic and prescribe metronidazole 500 mg TID × 10–14 days.

- If there is no improvement after 2–3 days (based on severity), or diarrhea subsides and recurs, refer the patient to his or her family physician, who will evaluate fluid/electrolyte status. For severe cases, the physician may switch metronidazole to oral vancomycin, which is not absorbed but provides its action locally within the colon. However, oral vancomycin is shockingly expensive and will be initiated only in extreme cases.

It is significant that C difficile infection has been reported to occur as late as 6–8 weeks following clindamycin use and has also been associated with the use of macrolides, tetracycline, and most of the broader-spectrum beta lactam antibiotics. This complication is unheard of following abbreviated use of clindamycin for prophylaxis of infective endocarditis.

The use of probiotics to prevent or manage antibiotic-associated diarrhea remains controversial. Nevertheless, current evidence suggests they are indeed effective and should be suggested for particularly frail patients or those who have experienced diarrhea with antibiotic regimens in the past.

Antibiotic Drug Interactions

Like all drug classes, antibiotics pose a risk for drug interactions. Since the early 1980s, numerous articles have suggested that antibiotics may reduce the efficacy of oral contraceptives (OCs). However, most of these publications have been either anecdotal reports or elaborate theories based on these reports. Unfortunately, many of these are found in dental literature. Anonymous cases are frequently cited in these careless reports and mention dentists who were sued for child support following an unwanted pregnancy attributed to antibiotic-OC interaction. None of these case reports cite legal proceedings that can be researched, and these reports should be viewed with suspicion. To date, rifampin (an antituberculosis agent) is the only antibiotic having a confirmed interaction with OCs.

In a thorough review of OCs, Sondheimer stated that the generally accepted rate of unwanted pregnancy among OC users is 0.5–1% (8% among teens) and the most common reason for failure is noncompliance. Furthermore, he added that most antibiotics do not decrease the effectiveness of the birth control pill. It is entirely within reason to suspect that a woman might also be taking an antibiotic during the month of OC failure, most likely penicillin or tetracycline. To date, all human studies measuring the influence of antibiotics on estrogen and progestin serum concentrations have found no interaction with antibiotics other than rifampin. There is simply no sound evidence to support the contention that antibiotics other than rifampin reduce the effectiveness of OCs.

Fortunately, the beta lactam antibiotics used to manage dental infections are virtually free of any significant drug interactions. The macrolides are an entirely different matter, however. They not only have an inherent potential to lengthen cardiac repolarization and prolong QT intervals but inhibit the metabolism of many drugs having even more potential for this cardiotoxic effect. They also delay metabolism and elevate serum concentrations of additional drugs, leading to potential toxicity. These include warfarin, digoxin, carbamazepine, valproic acid, ergotamine, and theophyllines. This alarming list of potential drug interactions, combined with their declining activity against dental pathogens, renders the routine use of macrolides questionable.

The potential for interactions with warfarin deserves particular mention. Along with macrolides, metronidazole and azole antifungals delay the metabolism of warfarin and must always be avoided. Other antibiotics have some potential for enhancing warfarin effects by reducing bowel flora that produce vitamin K, but this potential is generally not a concern. Nonetheless, caution is advised when caring for older patients. A recent publication found that exposure to several classes of antimicrobials, particularly azole antifungals, was associated with an increased risk of bleeding in this population receiving warfarin.

Finally, doxycycline may elevate digoxin levels and should be avoided. It has also been found to enhance the hypoglycemic influences of exogenously administered insulin. Diabetics receiving insulin should be cautioned to monitor their glucose closely if doxycycline is prescribed.

ANTIVIRAL DRUG

Antiviral drugs are a class of medication used specifically for treating viral infections rather than bacterial ones. Most antivirals are used for specific viral infections, while a broad-spectrum antiviral is effective against a wide range of viruses. Unlike most antibiotics, antiviral drugs do not destroy their target pathogen; instead they inhibit their development.

Antiviral drugs are one class of antimicrobials, a larger group which also includes antibiotic (also termed antibacterial), antifungal and antiparasitic drugs, or antiviral drugs based on monoclonal antibodies. Most antivirals are considered relatively harmless to the host, and therefore can be used to treat infections. They should be distinguished from viricides, which are not medication but deactivate or destroy virus particles, either inside or outside the body. Natural antivirals are produced by some plants such as eucalyptus and Australian tea trees.

Medical Uses

Most of the antiviral drugs now available are designed to help deal with HIV, herpes viruses, the hepatitis B and C viruses, and influenza A and B viruses. Researchers are working to extend the range of antivirals to other families of pathogens.

Designing safe and effective antiviral drugs is difficult, because viruses use the host's cells to replicate. This makes it difficult to find targets for the drug that would interfere with the virus without also harming the host organism's cells. Moreover, the major difficulty in developing vaccines and anti-viral drugs is due to viral variation.

The emergence of antivirals is the product of a greatly expanded knowledge of the genetic and molecular function of organisms, allowing biomedical researchers to understand the structure and function of viruses, major advances in the techniques for finding new drugs, and the pressure placed on the medical profession to deal with the human immunodeficiency virus (HIV), the cause of acquired immunodeficiency syndrome (AIDS).

The first experimental antivirals were developed in the 1960s, mostly to deal with herpes viruses, and were found using traditional trial-and-error drug discovery methods. Researchers grew cultures of cells and infected them with the target virus. They then introduced into the cultures chemicals which they thought might inhibit viral activity, and observed whether the level of virus in the cultures rose or fell. Chemicals that seemed to have an effect were selected for closer study.

This was a very time-consuming, hit-or-miss procedure, and in the absence of a good knowledge of how the target virus worked, it was not efficient in discovering effective antivirals which had few side effects. Only in the 1980s, when the full genetic sequences of viruses began to be unraveled, did researchers begin to learn how viruses worked in detail, and exactly what chemicals were needed to thwart their reproductive cycle.

Virus Life Cycle

Viruses consist of a genome and sometimes a few enzymes stored in a capsule made of protein (called a capsid), and sometimes covered with a lipid layer (sometimes called an 'envelope'). Viruses cannot reproduce on their own, and instead propagate by subjugating a host cell to produce copies of themselves, thus producing the next generation.

Researchers working on such "rational drug design" strategies for developing antivirals have tried to attack viruses at every stage of their life cycles. Some species of mushrooms have been found to contain multiple antiviral chemicals with similar synergistic effects. Compounds Isolated from fruiting bodies and filtrates of various mushrooms have broad spectrum antiviral activities, but successful production and availability of such compounds as frontline antiviral is a long way away. Viral life cycles vary in their precise details depending on the type of virus, but they all share a general pattern:

- Attachment to a host cell.
- Release of viral genes and possibly enzymes into the host cell.
- Replication of viral components using host-cell machinery.
- Assembly of viral components into complete viral particles.
- Release of viral particles to infect new host cells.

Anti Viral Targeting

The general idea behind modern antiviral drug design is to identify viral proteins, or parts of proteins, that can be disabled. These "targets" should generally be as unlike any proteins or parts of proteins in humans as possible, to reduce the likelihood of side effects. The targets should also be common across many strains of a virus, or even among different species of virus in the same family, so a single drug will have broad effectiveness. For example, a researcher might target a critical enzyme synthesized by the virus, but not the patient, that is common across strains, and see what can be done to interfere with its operation.

Once targets are identified, candidate drugs can be selected, either from drugs already known to have appropriate effects, or by actually designing the candidate at the molecular level with a computer-aided design program.

The target proteins can be manufactured in the lab for testing with candidate treatments by inserting the gene that synthesizes the target protein into bacteria or other kinds of cells. The cells are then cultured for mass production of the protein, which can then be exposed to various treatment candidates and evaluated with "rapid screening" technologies.

Approaches by Life Cycle Stage

Before Cell Entry

One anti-viral strategy is to interfere with the ability of a virus to infiltrate a target cell. The virus must go through a sequence of steps to do this, beginning with binding to a specific "receptor" molecule on the surface of the host cell and ending with the virus "uncoating" inside the cell and releasing its contents. Viruses that have a lipid envelope must also fuse their envelope with the target cell, or with a vesicle that transports them into the cell, before they can uncoat.

This stage of viral replication can be inhibited in two ways:

- Using agents which mimic the virus-associated protein (VAP) and bind to the cellular receptors. This may include VAP anti-idiotypic antibodies, natural ligands of the receptor and anti-receptor antibodies.
- Using agents which mimic the cellular receptor and bind to the VAP. This includes anti-VAP antibodies, receptor anti-idiotypic antibodies, extraneous receptor and synthetic receptor mimics.

This strategy of designing drugs can be very expensive, and since the process of generating anti-idiotypic antibodies is partly trial and error, it can be a relatively slow process until an adequate molecule is produced.

Entry Inhibitor

A very early stage of viral infection is viral entry, when the virus attaches to and enters the host cell. A number of "entry-inhibiting" or "entry-blocking" drugs are being developed to fight HIV. HIV most heavily targets the immune system's white blood cells known as "helper T cells", and identifies these target cells through T-cell surface receptors designated "CD4" and "CCR5". Attempts to interfere with the binding of HIV with the CD4 receptor have failed to stop HIV from infecting helper T cells, but research continues on trying to interfere with the binding of HIV to the CCR5 receptor in hopes that it will be more effective.

HIV infects a cell through fusion with the cell membrane, which requires two different cellular molecular participants, CD4 and a chemokine receptor (differing depending on the cell type). Approaches to blocking this virus/cell fusion have shown some promise in preventing entry of the virus into a cell. At least one of these entry inhibitors—a biomimetic peptide called Enfuvirtide, or the brand name Fuzeon—has received FDA approval and has been in use for some time. Potentially, one of the benefits from the use of an effective entry-blocking or entry-inhibiting agent is that it potentially may not only prevent the spread of the virus within an infected individual but also the spread from an infected to an uninfected individual.

One possible advantage of the therapeutic approach of blocking viral entry (as opposed to the currently dominant approach of viral enzyme inhibition) is that it may prove more difficult for the virus to develop resistance to this therapy than for the virus to mutate or evolve its enzymatic protocols.

Uncoating Inhibitor

Inhibitors of uncoating have also been investigated. Amantadine and rimantadine have been introduced to combat influenza. These agents act on penetration and uncoating.

Pleconaril works against rhinoviruses, which cause the common cold, by blocking a pocket on the surface of the virus that controls the uncoating process. This pocket is similar in most strains of rhinoviruses and enteroviruses, which can cause diarrhea, meningitis, conjunctivitis, and encephalitis.

Some scientists are making the case that a vaccine against rhinoviruses, the predominant cause of the common cold, is achievable. Vaccines that combine dozens of varieties of rhinovirus at once are effective in stimulating antiviral antibodies in mice and monkeys, researchers have reported in Nature Communications in 2016.

Rhinoviruses are the most common cause of the common cold; other viruses such as respiratory syncytial virus, parainfluenza virus and adenoviruses can cause them too. Rhinoviruses also exacerbate asthma attacks. Although rhinoviruses come in many varieties, they do not drift to the same degree that influenza viruses do. A mixture of 50 inactivated rhinovirus types should be able to stimulate neutralizing antibodies against all of them to some degree.

During Viral Synthesis

A second approach is to target the processes that synthesize virus components after a virus invades a cell.

Reverse Transcription

One way of doing this is to develop nucleotide or nucleoside analogues that look like the building blocks of RNA or DNA, but deactivate the enzymes that synthesize the RNA or DNA once the analogue is incorporated. This approach is more commonly associated with the inhibition of reverse transcriptase (RNA to DNA) than with "normal" transcriptase (DNA to RNA).

The first successful antiviral, aciclovir, is a nucleoside analogue, and is effective against herpesvirus infections. The first antiviral drug to be approved for treating HIV, zidovudine (AZT), is also a nucleoside analogue.

An improved knowledge of the action of reverse transcriptase has led to better nucleoside analogues to treat HIV infections. One of these drugs, lamivudine, has been approved to treat hepatitis B, which uses reverse transcriptase as part of its replication process. Researchers have gone further and developed inhibitors that do not look like nucleosides, but can still block reverse transcriptase.

Another target being considered for HIV antivirals include RNase H – which is a component of reverse transcriptase that splits the synthesized DNA from the original viral RNA.

Integrase

Another target is integrase, which integrate the synthesized DNA into the host cell genome.

Transcription

Once a virus genome becomes operational in a host cell, it then generates messenger RNA (mRNA) molecules that direct the synthesis of viral proteins. Production of mRNA is initiated by proteins known as transcription factors. Several antivirals are now being designed to block attachment of transcription factors to viral DNA.

Translation/Antisense

Genomics has not only helped find targets for many antivirals, it has provided the basis for an entirely new type of drug, based on "antisense" molecules. These are segments of DNA or RNA that are designed as complementary molecule to critical sections of viral genomes, and the binding of these antisense segments to these target sections blocks the operation of those genomes. A phosphorothioate antisense drug named fomivirsen has been introduced, used to treat opportunistic eye infections in AIDS patients caused by cytomegalovirus, and other antisense antivirals are in development. An antisense structural type that has proven especially valuable in research is morpholino antisense.

Morpholino oligos have been used to experimentally suppress many viral types:

- Caliciviruses.
- Flaviviruses (including WNV).
- Dengue.
- HCV.
- Coronaviruses.

Translation/Ribozymes

Yet another antiviral technique inspired by genomics is a set of drugs based on ribozymes, which are enzymes that will cut apart viral RNA or DNA at selected sites. In their natural course, ribozymes are used as part of the viral manufacturing sequence, but these synthetic ribozymes are designed to cut RNA and DNA at sites that will disable them.

A ribozyme antiviral to deal with hepatitis C has been suggested, and ribozyme antivirals are being developed to deal with HIV. An interesting variation of this idea is the use of genetically modified cells that can produce custom-tailored ribozymes. This is part of a broader effort to create genetically modified cells that can be injected into a host to attack pathogens by generating specialized proteins that block viral replication at various phases of the viral life cycle.

Protein Processing and Targeting

Interference with post translational modifications or with targeting of viral proteins in the cell is also possible.

Protease Inhibitors

Some viruses include an enzyme known as a protease that cuts viral protein chains apart so they can be assembled into their final configuration. HIV includes a protease, and so considerable research has been performed to find "protease inhibitors" to attack HIV at that phase of its life cycle. Protease inhibitors became available in the 1990s and have proven effective, though they can have unusual side effects, for example causing fat to build up in unusual places. Improved protease inhibitors are now in development.

Protease inhibitors have also been seen in nature. A protease inhibitor was isolated from the Shiitake mushroom (*Lentinus edodes*). The presence of this may explain the Shiitake mushroom's noted antiviral activity *in vitro*.

Assembly

Rifampicin acts at the assembly phase.

Release Phase

The final stage in the life cycle of a virus is the release of completed viruses from the host cell, and this step has also been targeted by antiviral drug developers. Two drugs named zanamivir (Relenza) and oseltamivir (Tamiflu) that have been recently introduced to treat influenza prevent the release of viral particles by blocking a molecule named neuraminidase that is found on the surface of flu viruses, and also seems to be constant across a wide range of flu strains.

Immune System Stimulation

A second category of tactics for fighting viruses involves encouraging the body's immune system to attack them, rather than attacking them directly. Some antivirals of this sort do not focus on a specific pathogen, instead stimulating the immune system to attack a range of pathogens.

One of the best-known of this class of drugs are interferons, which inhibit viral synthesis in infected cells. One form of human interferon named "interferon alpha" is well-established as part of the standard treatment for hepatitis B and C, and other interferons are also being investigated as treatments for various diseases.

A more specific approach is to synthesize antibodies, protein molecules that can bind to a pathogen and mark it for attack by other elements of the immune system. Once researchers identify a particular target on the pathogen, they can synthesize quantities of identical "monoclonal" antibodies to link up that target. A monoclonal drug is now being sold to help fight respiratory syncytial virus in babies, and antibodies purified from infected individuals are also used as a treatment for hepatitis B.

Acquired Resistance

Antiviral resistance can be defined by a decreased susceptibility to a drug caused by changes in viral genotypes. In cases of antiviral resistance, drugs have either diminished or no effectiveness against their target virus. The issue inevitably remains a major obstacle to antiviral therapy as it has developed to almost all specific and effective antimicrobials, including antiviral agents.

The Centers for Disease Control and Prevention (CDC) inclusively recommends those six months and older to get a yearly vaccination to protect from influenza A viruses (H1N1) and (H3N2) and up to two influenza B viruses (depending on the vaccination). Comprehensive protection starts by ensuring vaccinations are current and complete. However, vaccines are preventative and are not generally used once a patient has been infected with a virus. Additionally, the availability of these vaccines can be limited based on financial or locational reasons which can prevent the effectiveness of herd immunity, making effective antivirals a necessity.

The three FDA-approved neuraminidase antiviral flu drugs available in the United States, recommended by the CDC, include: oseltamivir (Tamiflu), zanamivir (Relenza), and peramivir (Rapivab). Influenza antiviral resistance often results from changes occurring in neuraminidase and hemagglutinin proteins on the viral surface. Currently, neuraminidase inhibitors (NAIs) are the most frequently prescribed antivirals because they are effective against both influenza A and B. However, antiviral resistance is known to develop if mutations to the neuraminidase proteins prevent NAI binding. This was seen in the H257Y mutation, which was responsible for oseltamivir resistance to H1N1 strains in 2009. The inability of NA inhibitors to bind to the virus allowed this strain of virus with the resistance mutation to spread due to natural selection. Furthermore, a study published in 2009 in Nature Biotechnology emphasized the urgent need for augmentation of oseltamivir (Tamiflu) stockpiles with additional antiviral drugs including zanamivir (Relenza). This finding was based on a performance evaluation of these drugs supposing the 2009 H1N1 'Swine Flu' neuraminidase (NA) were to acquire the Tamiflu-resistance (His274Tyr) mutation which is currently widespread in seasonal H1N1 strains.

Origin of Antiviral Resistance

The genetic makeup of viruses is constantly changing, which can cause a virus to become resistant to currently available treatments. Viruses can become resistant through spontaneous or intermittent mechanisms throughout the course of an antiviral treatment. Immunocompromised patients, more often than immunocompetent patients, hospitalized with pneumonia are at the highest risk of developing oseltamivir resistance during treatment. Subsequent to exposure to someone else with the flu, those who received oseltamivir for "post-exposure prophylaxis" are also at higher risk of resistance.

The mechanisms for antiviral resistance development depend on the type of virus in question. RNA viruses such as hepatitis C and influenza A have high error rates during genome replication because RNA polymerases lack proofreading activity. RNA viruses also have small genome sizes that are typically less than 30 kb, which allow them to sustain a high frequency of mutations. DNA viruses, such as HPV and herpesvirus, hijack host cell replication machinery, which gives them proofreading capabilities during replication. DNA viruses are therefore less error prone, are generally less diverse, and are more slowly evolving than RNA viruses. In both cases, the likelihood of mutations is exacerbated by the speed with which viruses reproduce, which provides more opportunities for mutations to occur in successive replications. Billions of viruses are produced every day during the course of an infection, with each replication giving another chance for mutations that encode for resistance to occur.

Multiple strains of one virus can be present in the body at one time, and some of these strains may contain mutations that cause antibiotic resistance. This effect, called the quasispecies model, results in immense variation in any given sample of virus, and gives the opportunity for natural selection to favor viral strains with the highest fitness every time the virus is spread to a new host. Also, recombination, the joining of two different viral variants, and reassortment, the swapping of viral gene segments among viruses in the same cell, play a role in resistance, especially in influenza.

Antiviral resistance has been reported in antivirals for herpes, HIV, hepatitis B and C, and influenza, but antiviral resistance is a possibility for all viruses. Mechanisms of antiviral resistance vary between virus types.

Detection of Antiviral Resistance

National and international surveillance is performed by the CDC to determine effectiveness of the current FDA-approved antiviral flu drugs. Public health officials use this information to make current recommendations about the use of flu antiviral medications. WHO further recommends in-depth epidemiological investigations to control potential transmission of the resistant virus and prevent future progression. As novel treatments and detection techniques to antiviral resistance are enhanced so can the establishment of strategies to combat the inevitable emergence of antiviral resistance.

Treatment Options for Antiviral Resistant Pathogens

If a virus is not fully wiped out during a regimen of antivirals, treatment creates a bottleneck in the viral population that selects for resistance, and there is a chance that a resistant strain may repopulate the host. Viral treatment mechanisms must therefore account for the selection of resistant viruses.

The most commonly used method for treating resistant viruses is combination therapy, which uses multiple antivirals in one treatment regimen. This is thought to decrease the likelihood that one mutation could cause antiviral resistance, as the antivirals in the cocktail target different stages of the viral life cycle. This is frequently used in retroviruses like HIV, but a number of studies have demonstrated its effectiveness against influenza A, as well. Viruses can also be screened for resistance to drugs before treatment is started. This minimizes exposure to unnecessary antivirals and ensures that an effective medication is being used. This may improve patient outcomes and could help detect new resistance mutations during routine scanning for known mutants. However, this has not been consistently implemented in treatment facilities at this time.

Oseltamivir

Oseltamivir, sold under the brand name Tamiflu, is an antiviral medication used to treat and prevent influenza A and influenza B (flu). Many medical organizations recommend it in people who have complications or are at high risk of complications within 48 hours of first symptoms of infection. They recommend it to prevent infection in those at high risk, but not the general population. The CDC recommends that clinicians use their discretion to treat those at lower risk who present within 48 hours of first symptoms of infection. It is taken by mouth, either as a pill or liquid.

Recommendations regarding oseltamivir are controversial as are criticisms of the recommendations. A 2014 Cochrane review concluded that oseltamivir does not reduce hospitalizations, and that there is no evidence of reduction in complications of influenza. Two meta-analyses have concluded that benefits in those who are otherwise healthy do not outweigh its risks. They also found little evidence regarding whether treatment changes the risk of hospitalization or death in high risk populations. However, another meta-analysis found that oseltamivir was effective for prevention of influenza at the individual and household levels.

Common side effects include vomiting, diarrhea, headache, and trouble sleeping. Other side effects may include psychiatric symptoms and seizures. In the United States it is recommended for influenza infection during pregnancy. It has been taken by a small number of pregnant women without signs of problems. Dose adjustment may be needed in those with kidney problems.

Oseltamivir was approved for medical use in the US in 1999. It was the first neuraminidase inhibitor available by mouth. It is on the complementary list of World Health Organization's List of Essential Medicines, indicating a lower cost-benefit ratio. A generic version was approved in the US in 2016. As of 2014 the wholesale cost in the developing world was about US$4.27 per day. The wholesale cost for a course of treatment in the United States is about US$54.00 as of 2019. In 2016 it was the 249th most prescribed medication in the United States with more than a million prescriptions.

Side Effects

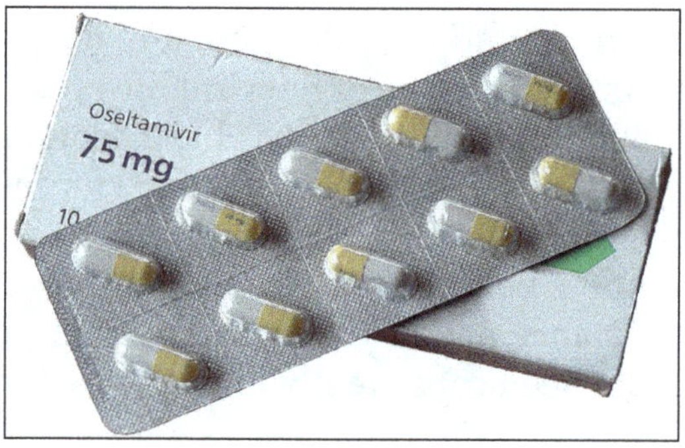

A package of capsules.

Common adverse drug reactions (ADRs) associated with oseltamivir therapy (occurring in over 1 percent of people) include nausea and vomiting. In adults, oseltamivir increased the risk of nausea for which the number needed to harm was 28 and for vomiting was 22. So, for every 22 adult people on oseltamivir one experienced vomiting. In the treatment of children, oseltamivir also induced vomiting. The number needed to harm was 19. So, for every 19 children on oseltamivir one experienced vomiting. In prevention there were more headaches, kidney, and psychiatric events. Oseltamivir's effect on the heart is unclear: it may reduce cardiac symptoms, but may also induce serious heart rhythm problems.

Postmarketing reports include liver inflammation and elevated liver enzymes, rash, allergic reactions including anaphylaxis, toxic epidermal necrolysis, abnormal heart rhythms, seizure, confusion, aggravation of diabetes, and haemorrhagic colitis and Stevens–Johnson syndrome.

The US and EU package inserts for oseltamivir contain a warning of psychiatric effects observed in post-marketing surveillance. The frequency of these appears to be low and a causative role for oseltamivir has not been established. The 2014 Cochrane review found a dose-response effect on psychiatric events. In trials of prevention in adults one person was harmed for every 94 treated. Neither of the two most cited published treatment trials of oseltamivir reported any drug-attributable serious adverse events.

It is pregnancy category C in the United States and category B in Australia, meaning that it has been taken by a small number of women without signs of problems and in animal studies it looks safe. Dose adjustment may be needed in those with kidney problems.

Mechanism of Action

Oseltamivir is a neuraminidase inhibitor, a competitive inhibitor of influenza's neuraminidase enzyme. The enzyme cleaves the sialic acid which is found on glycoproteins on the surface of human cells that helps new virions to exit the cell. Thus oseltamivir prevents new viral particles from being released.

Resistance

The vast majority of mutations conferring resistance are single amino acid residue substitutions (His274Tyr in N1) in the neuraminidase enzyme. A 2011 meta-analysis of 15 studies found a pooled incidence rate for oseltamivir resistance of 2.6%. Subgroup analyses detected higher rates among influenza A patients, especially the H1N1 subtype. It was found that a substantial number of patients might become oseltamivir-resistant as a result of oseltamivir use, and that oseltamivir resistance might be significantly associated with pneumonia. In severely immunocompromised patients there were reports of prolonged shedding of oseltamivir-(or zanamivir)-resistant virus, even after oseltamivir treatment was stopped.

H1N1 Flu or Swine Flu

As of December 15, 2010, the World Health Organization (WHO) reported 314 samples of the prevalent 2009 pandemic H1N1 flu tested worldwide have shown resistance to oseltamivir.

The CDC found sporadic oseltamivir-resistant 2009 H1N1 virus infections had been identified, including with rare episodes of limited transmission, but the public health impact had been limited. Those sporadic cases of resistance were found in immunosuppressed patients during oseltamivir treatment and persons who developed illness while receiving oseltamivir chemoprophylaxis.

During 2011 a new influenza A(H1N1)2009 variant with mildly reduced oseltamivir (and zanamivir) sensitivity was detected in more than 10% of community specimens in Singapore and more than 30% of samples from northern Australia.

While there is concern that antiviral resistance may develop in people with haematologic malignancies due to their inability to reduce viral loads and several surveillance studies found oseltamivir-resistant pH1N1 after administration of oseltamivir in those people, as of November 2013 widespread transmission of oseltamivir-resistant pH1N1 has not occurred.

During the 2007–08 flu season, the US CDC found 10.9% of H1N1 samples (n=1,020) to be resistant. In the 2008–09 season, the proportion of resistant H1N1 increased to 99.4%, while no other seasonal strains (H3N2, B) showed resistance.

Seasonal Flu

From 2009 to 2014 oseltamivir resistance was very low in seasonal flu. In the 2010–11 flu season, 99.1% of H1N1, 99.8% of H3N, and 100% of Influenza B remained oseltamivir susceptible in the US. In January 2012, the US and European CDCs reported all seasonal flu samples tested since October 2011 to be oseltamivir susceptible. In the 2013–14 season only 1% of 2009 H1N1 viruses showed oseltamivir resistance. No other influenza viruses were resistant to oseltamivir.

H3N2

Three studies have found resistance in 0%, 3.3%, and 18% of subjects. In the study with the 18% resistance rate, the subjects were children, many of whom had not been previously exposed to influenza and therefore had a weakened immune response; the results suggest that higher and earlier dosing may be necessary in such populations.

Influenza B

In 2007, Japanese investigators detected neuraminidase-resistant influenza B virus strains in individuals not treated with these drugs. The prevalence was 1.7 percent. According to the CDC, as of October 3, 2009 no influenza B strains tested had shown any resistance to oseltamivir.

H5N1 Avian Influenza Bird Flu

As of 2013 H274Y and N294S mutations that confer resistance to oseltamivir have been identified in a few H5N1 isolates from infected patients treated with oseltamivir, and have emerged spontaneously in Egypt.

H7N9 Avian Influenza

As of 2013 two of 14 adults infected with A(H7N9) and treated with oseltamivir developed oseltamivir-resistant virus with the Arg292Lys mutation.

Pharmacokinetics

Its oral bioavailability is over 80% and is extensively metabolised to its active form upon first-pass through the liver. It has a volume of distribution of 23–26 litres. Its half-life is about 1–3 hours and its active carboxylate metabolite has a half-life of 6–10 hours. More than 90% of the oral dose is eliminated in the urine as the active metabolite.

Veterinary Use

There have been reports of oseltamivir reducing disease severity and hospitalization time in canine parvovirus infection. The drug may limit the ability of the virus to invade the crypt cells of the small intestine and decrease gastrointestinal bacterial colonization and toxin production.

ANTIFUNGAL MEDICATION

An antifungal medication, also known as an antimycotic medication, is a pharmaceutical fungicide or fungistatic used to treat and prevent mycosis such as athlete's foot, ringworm, candidiasis (thrush), serious systemic infections such as cryptococcal meningitis, and others. Such drugs are usually obtained by a doctor's prescription, but a few are available OTC (over-the-counter).

Classes

Polyene Antifungals

A polyene is a molecule with multiple conjugated double bonds. A polyene antifungal is a macrocyclic polyene with a heavily hydroxylated region on the ring opposite the conjugated system. This makes polyene antifungals amphiphilic. The polyene antimycotics bind with sterols in the fungal cell membrane, principally ergosterol. This changes the transition temperature (Tg) of the cell membrane, thereby placing the membrane in a less fluid, more crystalline state. (In ordinary circumstances membrane sterols increase the packing of the phospholipid bilayer making the plasma membrane more dense.) As a result, the cell's contents including monovalent ions (K^+, Na^+, H^+, and Cl^-), small organic molecules leak and this is regarded one of the primary ways cell dies. Animal cells contain cholesterol instead of ergosterol and so they are much less susceptible. However, at therapeutic doses, some amphotericin B may bind to animal membrane cholesterol, increasing the risk of human toxicity. Amphotericin B is nephrotoxic when given intravenously. As a polyene's hydrophobic chain is shortened, its sterol binding activity is increased. Therefore, further reduction of the hydrophobic chain may result in it binding to cholesterol, making it toxic to animals.

- Amphotericin B.
- Candicidin.

- Filipin – 35 carbons, binds to cholesterol (toxic).
- Hamycin.
- Natamycin – 33 carbons, binds well to ergosterol.
- Nystatin.
- Rimocidin.

Adverse Effects

Apart from side effects like altered estrogen levels and liver damage, many antifungal medicines can cause allergic reactions in people. For example, the azole group of drugs is known to have caused anaphylaxis.

There are also many drug interactions. Patients must read in detail the enclosed data sheet(s) of any medicine. For example, the azole antifungals such as ketoconazole or itraconazole can be both substrates and inhibitors of the P-glycoprotein, which (among other functions) excretes toxins and drugs into the intestines. Azole antifungals also are both substrates and inhibitors of the cytochrome P450 family CYP3A4, causing increased concentration when administering, for example, calcium channel blockers, immunosuppressants, chemotherapeutic drugs, benzodiazepines, tricyclic antidepressants, macrolides and SSRIs.

Before oral antifungal therapies are used to treat nail disease, a confirmation of the fungal infection should be made. Approximately half of suspected cases of fungal infection in nails have a non-fungal cause. The side effects of oral treatment are significant and people without an infection should not take these drugs.

Azoles are the group of anti fungals which act on the cell membrane of fungus. They inhibit the enzyme 14 - alpha-sterol demethylase, a microsomal CYP, which is required for biosynthesis of Ergosterol for cytoplasmic membrane. This leads to accumulation of 14-alpha-methylsterols resulting in impairment of function of certain membrane bound enzymes and disruption close packing of acyl chains of phospholipids, thus inhibiting growth of the fungi. Some azoles directly increase permeability of fungal cell membrane.

Terbinafine

Terbinafine, sold under the brand name Lamisil among others, is an antifungal medication used to treat pityriasis versicolor, fungal nail infections, and ringworm including jock itch and athlete's foot. It is either taken by mouth or applied to the skin as a cream or ointment. The cream and ointment are not effective for nail infections.

Common side effects when taken by mouth include nausea, diarrhea, headache, cough, rash, and elevated liver enzymes. Severe side effects include liver problems and allergic reactions. Liver injury is, however, unusual. Use during pregnancy is not typically recommended. The cream and ointment may result in itchiness but are generally well tolerated. Terbinafine is in the allylamines family of medications. It works by decreasing the ability of fungi to make sterols. It appears to result in fungal cell death.

Terbinafine was discovered in 1991. It is on the World Health Organization's List of Essential Medicines, the most effective and safe medicines needed in a health system. The wholesale cost in the developing world is about 2.20 USD for a 20 g tube. In the United States the price in 1999 was $547 for a 12-week course; this fell to $10 by 2015, after the patent had expired. In 2016 it was the 279th most prescribed medication in the United States with more than a million prescriptions.

Medical Uses

Terbinafine is mainly effective on the dermatophyte group of fungi. As a cream or powder, it is used topically for superficial skin infections such as jock itch (tinea cruris), athlete's foot (tinea pedis), and other types of ringworm (tinea corporis).

Tablets by mouth are often prescribed for the treatment of onychomycosis, a fungal nail infection, typically by a dermatophyte or *Candida* species. Fungal nail infections are located deep under the nail in the cuticle to which topically applied treatments are unable to penetrate in sufficient amounts. The tablets may, rarely, cause hepatotoxicity, so patients are warned of this and may be monitored with liver function tests. Alternatives to by mouth administration have been studied.

Terbinafine hydrochloride may induce or exacerbate subacute cutaneous lupus erythematosus. Persons with lupus erythematosus should first discuss possible risks with their doctor before initiation of therapy.

FDA Approval

Terbinafine first became available in Europe in 1991 and in the United States in 1996. The U.S. Food and Drug Administration has approved the first generic versions of prescription Lamisil (terbinafine hydrochloride) tablets. The remaining patent or exclusivity for Lamisil expired on June 30, 2007.

On September 28, 2007, the FDA stated that terbinafine is a new treatment approved for use by children age four and up. The antifungal granules can be sprinkled on a child's food to treat ringworm of the scalp, tinea capitis.

Side Effects

Many side effects and adverse drug reactions have been reported with oral terbinafine hydrochloride possibly due to its extensive biodistribution and the often extended durations involved in antifungal treatment (longer than two months). A comprehensive list of adverse events associated with terbinafine use includes:

- Gastrointestinal problems: Diarrhea, constipation, nausea, fullness, abdominal pain, indigestion, dyspepsia, gastritis, cholestasis, flatulence, altered stool colour, abdominal muscular pain.
- Central nervous system or neurological problems: Headaches, dizziness, vertigo, light-headedness, decreased concentration levels, paraesthesia (pins and needles).
- Hepatic problems: Raised liver enzyme levels, liver inflammation (hepatitis), liver damage, liver failure.
- Immune system problems: Decreased white blood cell counts including pancytopenia, leukopenia, lymphopenia, thrombocytopenia, agranulocytosis, and neutropenia, autoimmune reactions such as lupus erythematosus.
- Psychological problems: Depression, anxiety, insomnia, increased or unusual dream activity, malaise.
- Sensory problems: Complete loss of taste (ageusia), decreased taste (hypogeusia) and distorted taste (dysgeusia), often involving a metallic taste sensation and dry mouth, visual disturbances including blurred vision, green vision and double vision. In extremely rare cases, the loss or impairment of taste is permanent.
- Skin problems: Rashes, hives (urticaria), skin irritation, itching, jaundice, Stevens–Johnson syndrome.
- Other side effects: Fatigue, increased heart rate (tachycardia), hair loss (alopecia), decreased red blood cell count (anemia), muscle pain (myalgia), joint pain (arthralgia).

In 2015 physicians reported that a patient with an MTHFR enzyme mutation (specifically the C677T variant) had developed an adverse reaction to Lamisil (headache, fatigue, and dizziness). Genetic testing revealed the MTHFR C677T mutation. It was noted that Lamisil interferes with the methylation cycle and that this can cause side effects in individuals with the MTHFR C677T mutation.

Pharmacology

Terbinafine hydrochloride is a white fine crystalline powder that is freely soluble in methanol and dichloromethane, soluble in ethanol, and slightly soluble in water. It is highly hydrophobic and tends to accumulate in hair, skin, nails, and fatty tissue.

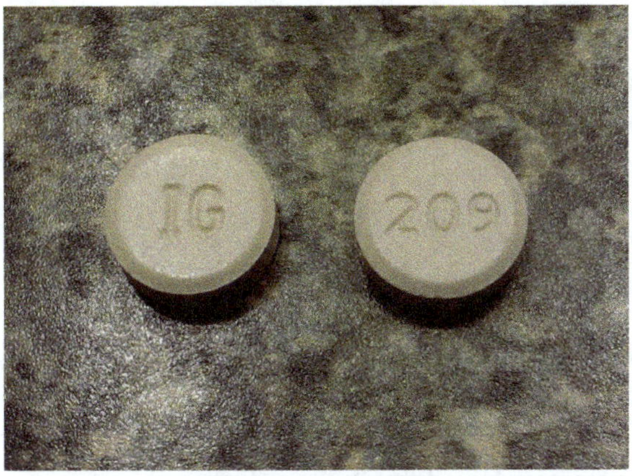

Generic terbinafine hydrochloride pills.

Like other allylamines, terbinafine inhibits ergosterol synthesis by inhibiting squalene epoxidase, an enzyme that catalyzes the conversion of squalene to lanosterol. In fungi, lanosterol is then converted to ergosterol; in humans, lanosterol becomes cholesterol. However, there is sufficient genetic divergence between fungal and human squalene epoxidases that terbinafine preferentially binds fungal squalene epoxidase, making it selective for inhibiting ergosterol production in fungi without significantly affecting cholesterol production in humans. This is thought to change cell membrane permeability, causing fungal cell lysis.

ANTIPARASITICS

Antiparasitics are a class of medications which are indicated for the treatment of parasitic diseases, such as those caused by helminths, amoeba, ectoparasites, parasitic fungi, and protozoa, among others. Antiparasitics target the parasitic agents of the infections by destroying them or inhibiting their growth; they are usually effective against a limited number of parasites within a particular class. Antiparasitics are one of the antimicrobial drugs which include antibiotics that target bacteria, and antifungals that target fungi. They may be administered orally, intravenously or topically.

Broad-Spectrum antiparasitics, analogous to broad-spectrum antibiotics for bacteria, are antiparasitic drugs with efficacy in treating a wide range of parasitic infections caused by parasites from different classes.

Types

Broad-spectrum

- Nitazoxanide.

Antiprotozoals

- Melarsoprol (for treatment of sleeping sickness caused by *Trypanosoma brucei*).
- Eflornithine (for sleeping sickness).
- Metronidazole (for vaginitis caused by Trichomonas).
- Tinidazole (for intestinal infections caused by *Giardia lamblia*).
- Miltefosine (for the treatment of visceral and cutaneous leishmaniasis, currently undergoing investigation for Chagas disease).

Antihelminthic

Antinematodes

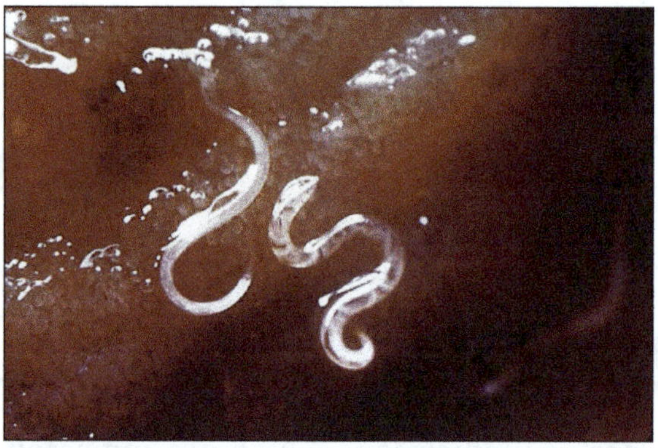

Ancylostoma caninum, a type of hookworm, attached to the intestinal mucosa.

- Mebendazole (for most nematode infections).
- Pyrantel pamoate (for most nematode infections).
- Thiabendazole (for roundworm infections).
- Diethylcarbamazine (for treatment of Lymphatic filariasis).
- Ivermectin (for prevention of river blindness).

Anticestodes

- Niclosamide (for tapeworm infections).
- Praziquantel (for tapeworm infections).
- Albendazole (broad spectrum).

Antitrematodes

- Praziquantel.

Antiamoebics

- Rifampin.
- Amphotericin B.

Antifungals

- Fumagillin (for microsporidiosis).

Medical Uses

Antiparasitics treat parasitic diseases, which impact an estimated 2 billion people.

Administration

Antiparastics may be given via a variety of routes depending on the specific medication, including oral, topical, and intravenous.

Resistance to antiparasitics has been growing concern, especially in veterinary medicine. The Egg hatch assay can be used to determine whether a parasite causing an infection has become resistant to standard drug treatments.

Antihelminthic

Anthelmintics or antihelminthics are a group of antiparasitic drugs that expel parasitic worms (helminths) and other internal parasites from the body by either stunning or killing them and without causing significant damage to the host. They may also be called vermifuges (those that stun) or vermicides (those that kill). Anthelmintics are used to treat people who are infected by helminths, a condition called helminthiasis. These drugs are also used to treat infected animals.

Pills containing anthelmintics are used in mass deworming campaigns of school-aged children in many developing countries. For example, the treatment of choice for soil-transmitted helminths is mebendazole and albendazole and praziquantel for schistosomiasis and tapeworms.

Types

Antiparasitics that specifically target *Ascaris* worms are called ascaricides.

- Benzimidazoles:
    - Albendazole: Effective against threadworms, roundworms, whipworms, tapeworms, hookworms.

    - Mebendazole: Effective against various nematodes.
    - Thiabendazole: Effective against various nematodes.
    - Fenbendazole: Effective against various parasites.
    - Triclabendazole: Effective against liver flukes.
    - Flubendazole: Effective against most intestinal parasites.
- Abamectin (and by extension ivermectin): Effective against most common intestinal worms, except tapeworms, for which praziquantel is commonly used in conjunction for mass dewormings.
- Diethylcarbamazine: Effective against *Wuchereria bancrofti*, *Brugia malayi*, *Brugia timori* and *Loa loa*.
- Pyrantel pamoate: Effective against most nematode infections residing within the intestines.
- Levamisole.
- Salicylanilide mitochondrial un-couplers (used only for flatworm infections):
    - Niclosamide.
    - Oxyclozanide.
- Nitazoxanide: Readily kills *Ascaris lumbricoides*, and also possess antiprotozoal effects.
- Praziquantel: Effective against flatworms (e.g., tapeworms and schistosoma).
- Octadepsipeptides (e.g.: Emodepside): effective against a variety of gastrointestinal helminths.
- Monepantel (aminoacetonitrile class): Effective against a variety of nematodes including those resistant to other anthelmintic classes.
- Spiroindoles (e.g., derquantel): Effective against a variety of nematodes including those resistant to other anthelmintic classes.
- Artemisinin shows anthelmintic activity.

Anthelmintic Resistance

The ability of parasites to survive treatments that are generally effective at the recommended doses is a major threat to the future control of worm parasites in small ruminants and horses. This is especially true of nematodes, and has helped spur development of aminoacetonitrile derivatives for treatment against drug-resistant nematodes, as well as exploration of doxycycline to kill their endosymbiotic Wolbachia bacteria.

The resistance is measured by the "fecal egg count reduction" value which varies for different types of helminths.

Treatment with an antihelminthic drug kills worms whose phenotype renders them susceptible to the drug, but resistant parasites survive and pass on their "resistance" genes. Resistant varieties accumulate and finally treatment failure occurs.

ANTIBIOTIC MEDICATION

An antibiotic is a type of antimicrobial substance active against bacteria and is the most important type of antibacterial agent for fighting bacterial infections. Antibiotic medications are widely used in the treatment and prevention of such infections. They may either kill or inhibit the growth of bacteria. A limited number of antibiotics also possess antiprotozoal activity. Antibiotics are not effective against viruses such as the common cold or influenza; drugs which inhibit viruses are termed antiviral drugs or antivirals rather than antibiotics.

The usual medical usage, antibiotics (such as penicillin) are those produced naturally (by one microorganism fighting another), whereas nonantibiotic antibacterials (such as sulfonamides and antiseptics) are fully synthetic. However, both classes have the same goal of killing or preventing the growth of microorganisms, and both are included in antimicrobial chemotherapy. "Antibacterials" include antiseptic drugs, antibacterial soaps, and chemical disinfectants, whereas antibiotics are an important class of antibacterials used more specifically in medicine and sometimes in livestock feed.

Antibiotics have been used since ancient times. Many civilizations used topical application of mouldy bread, with many references to its beneficial effects arising from ancient Egypt, China, Serbia, Greece and Rome. The first person to directly document the use of moulds to treat infections was John Parkinson. Antibiotics revolutionized medicine in the 20th century. Alexander Fleming discovered modern day penicillin in 1928. After realizing the great potential there was in penicillin, Fleming pursued the challenge of how to market it and translate it to commercial use. With help from other biochemists, penicillin was finally available for widespread use. This was significantly beneficial during wartime. Unfortunately, it didn't take long for resistance to begin. Effectiveness and easy access have also led to their overuse and some bacteria have developed resistance. This has led to widespread problems, and the World Health Organization has classified antimicrobial resistance as a "serious threat that is no longer a prediction for the future, it is happening right now in every region of the world and has the potential to affect anyone, of any age, in any country".

Medical Uses

Antibiotics are used to treat or prevent bacterial infections, and sometimes protozoan infections. (Metronidazole is effective against a number of parasitic diseases). When an infection is suspected of being responsible for an illness but the responsible pathogen has not been identified, an empiric therapy is adopted. This involves the administration of a broad-spectrum antibiotic based on the signs and symptoms presented and is initiated pending laboratory results that can take several days.

When the responsible pathogenic microorganism is already known or has been identified, definitive therapy can be started. This will usually involve the use of a narrow-spectrum antibiotic. The choice of antibiotic given will also be based on its cost. Identification is critically important as it can reduce the cost and toxicity of the antibiotic therapy and also reduce the possibility of the emergence of antimicrobial resistance. To avoid surgery, antibiotics may be given for non-complicated acute appendicitis.

Antibiotics may be given as a preventive measure and this is usually limited to at-risk populations such as those with a weakened immune system (particularly in HIV cases to prevent pneumonia), those taking immunosuppressive drugs, cancer patients, and those having surgery. Their use in surgical procedures is to help prevent infection of incisions. They have an important role in dental antibiotic prophylaxis where their use may prevent bacteremia and consequent infective endocarditis. Antibiotics are also used to prevent infection in cases of neutropenia particularly cancer-related.

Administration

There are many different routes of administration for antibiotic treatment. Antibiotics are usually taken by mouth. In more severe cases, particularly deep-seated systemic infections, antibiotics can be given intravenously or by injection. Where the site of infection is easily accessed, antibiotics may be given topically in the form of eye drops onto the conjunctiva for conjunctivitis or ear drops for ear infections and acute cases of swimmer's ear. Topical use is also one of the treatment options for some skin conditions including acne and cellulitis. Advantages of topical application include achieving high and sustained concentration of antibiotic at the site of infection; reducing the potential for systemic absorption and toxicity, and total volumes of antibiotic required are reduced, thereby also reducing the risk of antibiotic misuse. Topical antibiotics applied over certain types of surgical wounds have been reported to reduce the risk of surgical site infections. However, there are certain general causes for concern with topical administration of antibiotics. Some systemic absorption of the antibiotic may occur; the quantity of antibiotic applied is difficult to accurately dose, and there is also the possibility of local hypersensitivity reactions or contact dermatitis occurring.

Prevalence

Antibiotic consumption varies widely between countries. The 'WHO report on surveillance of antibiotic consumption' published in 2018 analysed 2015 data from 65 countries. As measured in defined daily doses per 1,000 inhabitants per day. Mongolia had the highest consumption with a rate of 64.4. Burundi had the lowest at 4.4. Amoxicillin and Amoxicillin/clavulanic acid were the most frequently consumed.

Side Effects

Antibiotics are screened for any negative effects before their approval for clinical use, and are usually considered safe and well tolerated. However, some antibiotics have been associated with a wide extent of adverse side effects ranging from mild to very severe depending on the type of antibiotic used, the microbes targeted, and the individual patient. Side effects may reflect the pharmacological or toxicological properties of the antibiotic or may involve hypersensitivity or allergic reactions. Adverse effects range from fever and nausea to major allergic reactions, including photodermatitis and anaphylaxis. Safety profiles of newer drugs are often not as well established as for those that have a long history of use.

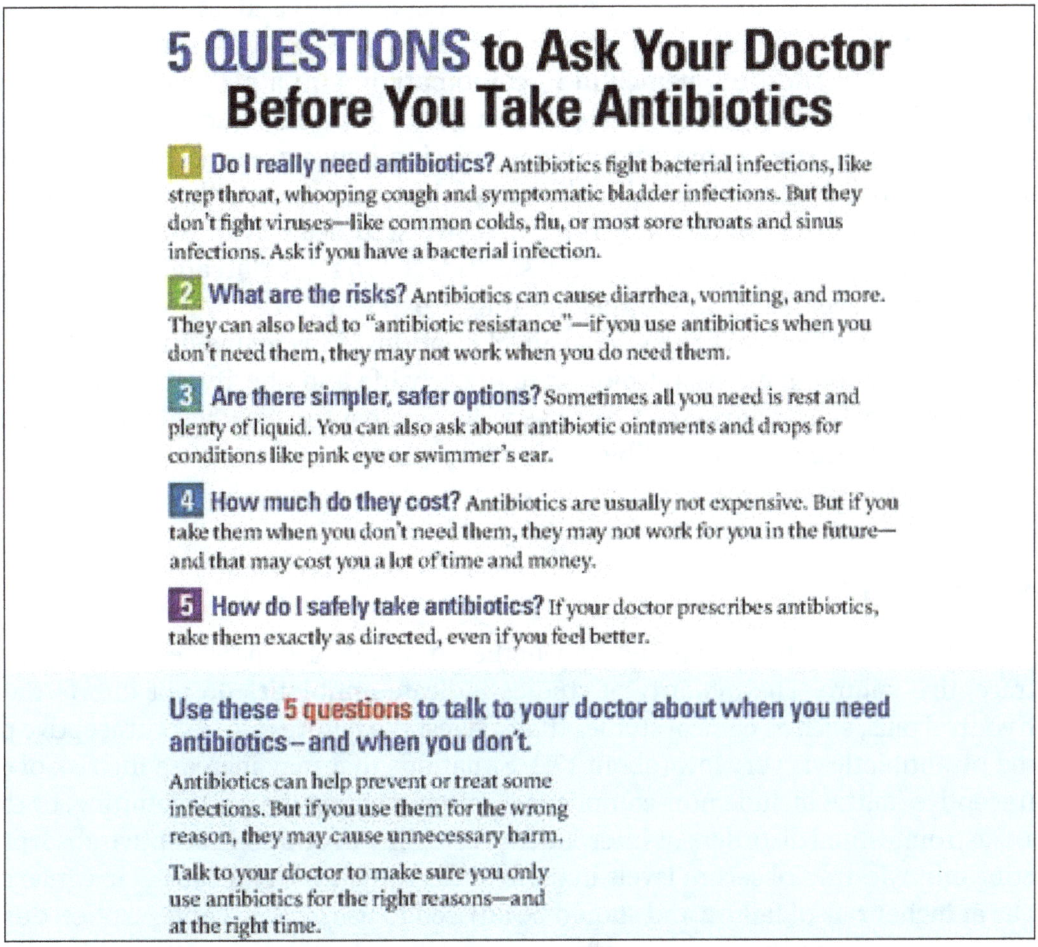

Health advocacy messages such as this one encourage patients to talk with their doctor about safety in using antibiotics.

Common side-effects include diarrhea, resulting from disruption of the species composition in the intestinal flora, resulting, for example, in overgrowth of pathogenic bacteria, such as *Clostridium difficile*. Antibacterials can also affect the vaginal flora, and may lead to overgrowth of yeast species of the genus *Candida* in the vulvo-vaginal area. Additional side effects can result from interaction with other drugs, such as the possibility of tendon damage from the administration of a quinolone antibiotic with a systemic corticosteroid.

Correlation with Obesity

Exposure to antibiotics early in life is associated with increased body mass in humans and mouse models. Early life is a critical period for the establishment of the intestinal microbiota and for metabolic development. Mice exposed to subtherapeutic antibiotic treatment (STAT)– with either penicillin, vancomycin, or chlortetracycline had altered composition of the gut microbiota as well as its metabolic capabilities. One study has reported that mice given low-dose penicillin (1 μg/g body weight) around birth and throughout the weaning process had an increased body mass and fat mass, accelerated growth, and increased hepatic expression of genes involved in adipogenesis, compared to control mice. In addition, penicillin in combination with a high-fat diet increased fasting insulin levels in mice. However, it is unclear whether or not antibiotics cause obesity in humans. Studies have found a correlation between early exposure of antibiotics (<6 months) and increased body mass (at 10 and 20 months). Another study found that the type of antibiotic exposure was also significant with the highest risk of being overweight in those given macrolides compared to penicillin and cephalosporin. Therefore, there is correlation between antibiotic exposure in early life and obesity in humans, but whether or not there is a causal relationship remains unclear. Although there is a correlation between antibiotic use in early life and obesity, the effect of antibiotics on obesity in humans needs to be weighed against the beneficial effects of clinically indicated treatment with antibiotics in infancy.

Interactions

Birth Control Pills

There are few well-controlled studies on whether antibiotic use increases the risk of oral contraceptive failure. The majority of studies indicate antibiotics do not interfere with birth control pills, such as clinical studies that suggest the failure rate of contraceptive pills caused by antibiotics is very low (about 1%). Situations that may increase the risk of oral contraceptive failure include non-compliance (missing taking the pill), vomiting, or diarrhea. Gastrointestinal disorders or interpatient variability in oral contraceptive absorption affecting ethinylestradiol serum levels in the blood. Women with menstrual irregularities may be at higher risk of failure and should be advised to use backup contraception during antibiotic treatment and for one week after its completion. If patient-specific risk factors for reduced oral contraceptive efficacy are suspected, backup contraception is recommended.

In cases where antibiotics have been suggested to affect the efficiency of birth control pills, such as for the broad-spectrum antibiotic rifampicin, these cases may be due to an increase in the activities of hepatic liver enzymes' causing increased breakdown of the pill's active ingredients. Effects on the intestinal flora, which might result in reduced absorption of estrogens in the colon, have also been suggested, but such suggestions have been inconclusive and controversial. Clinicians have recommended that extra contraceptive measures be applied during therapies using antibiotics that are suspected to interact with oral contraceptives. More studies on the possible interactions between antibiotics and birth control pills (oral contraceptives) are required as well as careful assessment of patient-specific risk factors for potential oral contractive pill failure prior to dismissing the need for backup contraception.

Alcohol

Interactions between alcohol and certain antibiotics may occur and may cause side effects and decreased effectiveness of antibiotic therapy. While moderate alcohol consumption is unlikely to interfere with many common antibiotics, there are specific types of antibiotics, with which alcohol consumption may cause serious side effects. Therefore, potential risks of side effects and effectiveness depend on the type of antibiotic administered.

Antibiotics such as metronidazole, tinidazole, cephamandole, latamoxef, cefoperazone, cefmenoxime, and furazolidone, cause a disulfiram-like chemical reaction with alcohol by inhibiting its breakdown by acetaldehyde dehydrogenase, which may result in vomiting, nausea, and shortness of breath. In addition, the efficacy of doxycycline and erythromycin succinate may be reduced by alcohol consumption. Other effects of alcohol on antibiotic activity include altered activity of the liver enzymes that break down the antibiotic compound.

Pharmacodynamics

The successful outcome of antimicrobial therapy with antibacterial compounds depends on several factors. These include host defense mechanisms, the location of infection, and the pharmacokinetic and pharmacodynamic properties of the antibacterial. A bactericidal activity of antibacterials may depend on the bacterial growth phase, and it often requires ongoing metabolic activity and division of bacterial cells. These findings are based on laboratory studies, and in clinical settings have also been shown to eliminate bacterial infection. Since the activity of antibacterials depends frequently on its concentration, *in vitro* characterization of antibacterial activity commonly includes the determination of the minimum inhibitory concentration and minimum bactericidal concentration of an antibacterial. To predict clinical outcome, the antimicrobial activity of an antibacterial is usually combined with its pharmacokinetic profile, and several pharmacological parameters are used as markers of drug efficacy.

Combination Therapy

In important infectious diseases, including tuberculosis, combination therapy (i.e., the concurrent application of two or more antibiotics) has been used to delay or prevent the emergence of resistance. In acute bacterial infections, antibiotics as part of combination therapy are prescribed for their synergistic effects to improve treatment outcome as the combined effect of both antibiotics is better than their individual effect. Methicillin-resistant Staphylococcus aureus infections may be treated with a combination therapy of fusidic acid and rifampicin. Antibiotics used in combination may also be antagonistic and the combined effects of the two antibiotics may be less than if the individual antibiotic was given as part of a monotherapy. For example, chloramphenicol and tetracyclines are antagonists to penicillins and U.S.s. However, this can vary depending on the species of bacteria. In general, combinations of a bacteriostatic antibiotic and bactericidal antibiotic are antagonistic.

Classes

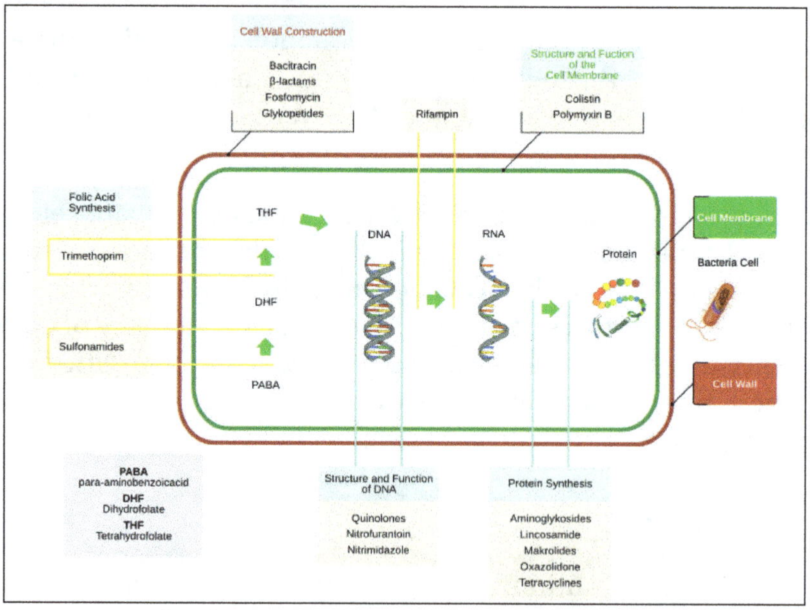

Molecular targets of antibiotics on the bacteria cell.

Antibiotics are commonly classified based on their mechanism of action, chemical structure, or spectrum of activity. Most target bacterial functions or growth processes. Those that target the bacterial cell wall (penicillins and cephalosporins) or the cell membrane (polymyxins), or interfere with essential bacterial enzymes (rifamycins, lipiarmycins, quinolones, and sulfonamides) have bactericidal activities. Protein synthesis inhibitors (macrolides, lincosamides, and tetracyclines) are usually bacteriostatic (with the exception of bactericidal aminoglycosides). Further categorization is based on their target specificity. "Narrow-spectrum" antibiotics target specific types of bacteria, such as gram-negative or gram-positive, whereas broad-spectrum antibiotics affect a wide range of bacteria. Following a 40-year break in discovering new classes of antibacterial compounds, four new classes of antibiotics have been brought into clinical use in the late 2000s and early 2010s: cyclic lipopeptides (such as daptomycin), glycylcyclines (such as tigecycline), oxazolidinones (such as linezolid), and lipiarmycins (such as fidaxomicin).

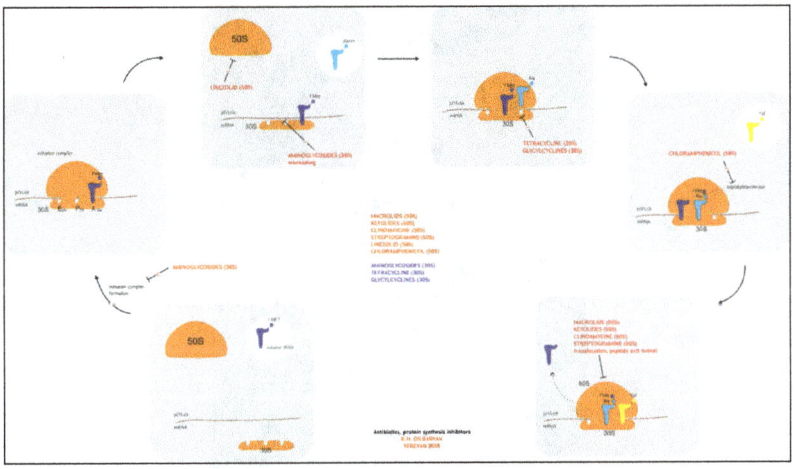

Protein synthesis inhibitors (antibiotics).

Production

With advances in medicinal chemistry, most modern antibacterials are semisynthetic modifications of various natural compounds. These include, for example, the beta-lactam antibiotics, which include the penicillins (produced by fungi in the genus *Penicillium*), the cephalosporins, and the carbapenems. Compounds that are still isolated from living organisms are the aminoglycosides, whereas other antibacterials—for example, the sulfonamides, the quinolones, and the oxazolidinones—are produced solely by chemical synthesis. Many antibacterial compounds are relatively small molecules with a molecular weight of less than 1000 daltons.

Since the first pioneering efforts of Howard Florey and Chain in 1939, the importance of antibiotics, including antibacterials, to medicine has led to intense research into producing antibacterials at large scales. Following screening of antibacterials against a wide range of bacteria, production of the active compounds is carried out using fermentation, usually in strongly aerobic conditions.

Resistance

The emergence of resistance of bacteria to antibiotics is a common phenomenon. Emergence of resistance often reflects evolutionary processes that take place during antibiotic therapy. The antibiotic treatment may select for bacterial strains with physiologically or genetically enhanced capacity to survive high doses of antibiotics. Under certain conditions, it may result in preferential growth of resistant bacteria, while growth of susceptible bacteria is inhibited by the drug. For example, antibacterial selection for strains having previously acquired antibacterial-resistance genes was demonstrated in 1943 by the Luria–Delbrück experiment. Antibiotics such as penicillin and erythromycin, which used to have a high efficacy against many bacterial species and strains, have become less effective, due to the increased resistance of many bacterial strains.

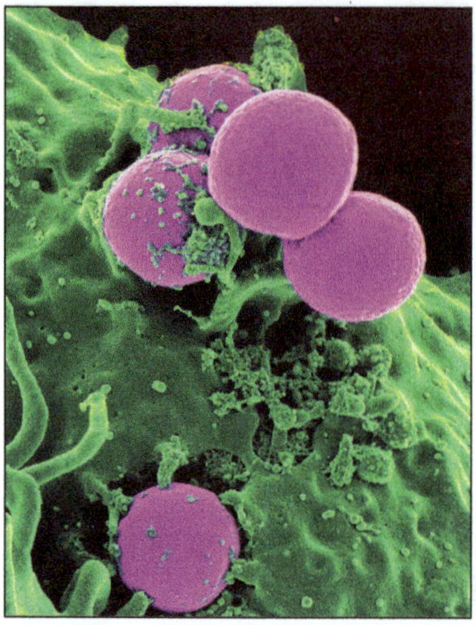

Scanning electron micrograph of a human neutrophil ingesting methicillin-resistant *Staphylococcus aureus* (MRSA).

Resistance may take the form of biodegredation of pharmaceuticals, such as sulfamethazine-degrading soil bacteria introduced to sulfamethazine through medicated pig feces. The survival of bacteria often results from an inheritable resistance, but the growth of resistance to antibacterials also occurs through horizontal gene transfer. Horizontal transfer is more likely to happen in locations of frequent antibiotic use.

Antibacterial resistance may impose a biological cost, thereby reducing fitness of resistant strains, which can limit the spread of antibacterial-resistant bacteria, for example, in the absence of antibacterial compounds. Additional mutations, however, may compensate for this fitness cost and can aid the survival of these bacteria.

Paleontological data show that both antibiotics and antibiotic resistance are ancient compounds and mechanisms. Useful antibiotic targets are those for which mutations negatively impact bacterial reproduction or viability.

Several molecular mechanisms of antibacterial resistance exist. Intrinsic antibacterial resistance may be part of the genetic makeup of bacterial strains. For example, an antibiotic target may be absent from the bacterial genome. Acquired resistance results from a mutation in the bacterial chromosome or the acquisition of extra-chromosomal DNA. Antibacterial-producing bacteria have evolved resistance mechanisms that have been shown to be similar to, and may have been transferred to, antibacterial-resistant strains. The spread of antibacterial resistance often occurs through vertical transmission of mutations during growth and by genetic recombination of DNA by horizontal genetic exchange. For instance, antibacterial resistance genes can be exchanged between different bacterial strains or species via plasmids that carry these resistance genes. Plasmids that carry several different resistance genes can confer resistance to multiple antibacterials. Cross-resistance to several antibacterials may also occur when a resistance mechanism encoded by a single gene conveys resistance to more than one antibacterial compound.

Antibacterial-resistant strains and species, sometimes referred to as "superbugs", now contribute to the emergence of diseases that were for a while well controlled. For example, emergent bacterial strains causing tuberculosis that are resistant to previously effective antibacterial treatments pose many therapeutic challenges. Every year, nearly half a million new cases of multidrug-resistant tuberculosis (MDR-TB) are estimated to occur worldwide. For example, NDM-1 is a newly identified enzyme conveying bacterial resistance to a broad range of beta-lactam antibacterials. The United Kingdom's Health Protection Agency has stated that "most isolates with NDM-1 enzyme are resistant to all standard intravenous antibiotics for treatment of severe infections." On 26 May 2016 an E coli bacteria "superbug" was identified in the United States resistant to colistin, "the last line of defence" antibiotic.

Misuse

This poster intended for use in doctors' offices and other healthcare facilities, warns that antibiotics do not work for viral illnesses such as the common cold.

"The first rule of antibiotics is try not to use them, and the second rule is try not to use too many of them." Inappropriate antibiotic treatment and overuse of antibiotics have contributed to the emergence of antibiotic-resistant bacteria. Self-prescribing of antibiotics is an example of misuse. Many antibiotics are frequently prescribed to treat symptoms or diseases that do not respond to antibiotics or that are likely to resolve without treatment. Also, incorrect or suboptimal antibiotics are prescribed for certain bacterial infections. The overuse of antibiotics, like penicillin and erythromycin, has been associated with emerging antibiotic resistance since the 1950s. Widespread usage of antibiotics in hospitals has also been associated with increases in bacterial strains and species that no longer respond to treatment with the most common antibiotics.

Common forms of antibiotic misuse include excessive use of prophylactic antibiotics in travelers and failure of medical professionals to prescribe the correct dosage of antibiotics on the basis of the patient's weight and history of prior use. Other forms of misuse include failure to take the entire prescribed course of the antibiotic, incorrect dosage and administration, or failure to rest for sufficient recovery. Inappropriate antibiotic treatment, for example, is their prescription to treat viral infections such as the common cold. One study on respiratory tract infections found "physicians were more likely to prescribe antibiotics to patients who appeared to expect them". Multifactorial interventions aimed at both physicians and patients can reduce inappropriate prescription of antibiotics. The lack of rapid point of care diagnostic tests, particularly in resource-limited settings is considered as one of the drivers of antibiotic misuse.

Several organizations concerned with antimicrobial resistance are lobbying to eliminate the unnecessary use of antibiotics. The issues of misuse and overuse of antibiotics have been addressed by the formation of the US Interagency Task Force on Antimicrobial Resistance. This task force aims to actively address antimicrobial resistance, and is coordinated by the US Centers for Disease Control and Prevention, the Food and Drug Administration (FDA), and the National Institutes of Health (NIH), as well as other US agencies. An NGO campaign group is *Keep Antibiotics Working*. In France, an "Antibiotics are not automatic" government campaign started in 2002 and led to a marked reduction of unnecessary antibiotic prescriptions, especially in children.

The emergence of antibiotic resistance has prompted restrictions on their use in the UK in 1970=, and the EU has banned the use of antibiotics as growth-promotional agents since 2003. Moreover, several organizations (including the World Health Organization, the National Academy of Sciences, and the U.S. Food and Drug Administration) have advocated restricting the amount of antibiotic use in food animal production. However, commonly there are delays in regulatory and legislative actions to limit the use of antibiotics, attributable partly to resistance against such regulation by industries using or selling antibiotics, and to the time required for research to test causal links between their use and resistance to them. Two federal bills (S.742 and H.R. 2562) aimed at phasing out nontherapeutic use of antibiotics in US food animals were proposed, but have not passed. These bills were endorsed by public health and medical organizations, including the American Holistic Nurses' Association, the American Medical Association, and the American Public Health Association (APHA).

Despite pledges by food companies and restaurants to reduce or eliminate meat that comes from animals treated with antibiotics, the purchase of antibiotics for use on farm animals has been increasing every year.

There has been extensive use of antibiotics in animal husbandry. In the United States, the question of emergence of antibiotic-resistant bacterial strains due to use of antibiotics in livestock was raised by the US Food and Drug Administration (FDA) in 1977. In March 2012, the United States District Court for the Southern District of New York, ruling in an action brought by the Natural Resources Defense Council and others, ordered the FDA to revoke approvals for the use of antibiotics in livestock, which violated FDA regulations.

Alternatives

The increase in bacterial strains that are resistant to conventional antibacterial therapies together with decreasing number of new antibiotics currently being developed in the drug pipeline has prompted the development of bacterial disease treatment strategies that are alternatives to conventional antibacterials. Non-compound approaches (that is, products other than classical antibacterial agents) that target bacteria or approaches that target the host including phage therapy and vaccines are also being investigated to combat the problem.

Resistance and Modifying Agents

One strategy to address bacterial drug resistance is the discovery and application of compounds that modify resistance to common antibacterials. Resistance modifying agents are capable of partly or completely suppressing bacterial resistance mechanisms. For example, some resistance-modifying agents may inhibit multidrug resistance mechanisms, such as drug efflux from the cell, thus increasing the susceptibility of bacteria to an antibacterial. Targets include:

- The efflux inhibitor Phe-Arg-β-naphthylamide.
- Beta-lactamase inhibitors, such as clavulanic acid and sulbactam.

Metabolic stimuli such as sugar can help eradicate a certain type of antibiotic-tolerant bacteria by keeping their metabolism active.

Vaccines

Vaccines rely on immune modulation or augmentation. Vaccination either excites or reinforces the immune competence of a host to ward off infection, leading to the activation of macrophages, the production of antibodies, inflammation, and other classic immune reactions. Antibacterial vaccines have been responsible for a drastic reduction in global bacterial diseases. Vaccines made from attenuated whole cells or lysates have been replaced largely by less reactogenic, cell-free vaccines consisting of purified components, including capsular polysaccharides and their conjugates, to protein carriers, as well as inactivated toxins (toxoids) and proteins.

Phage Therapy

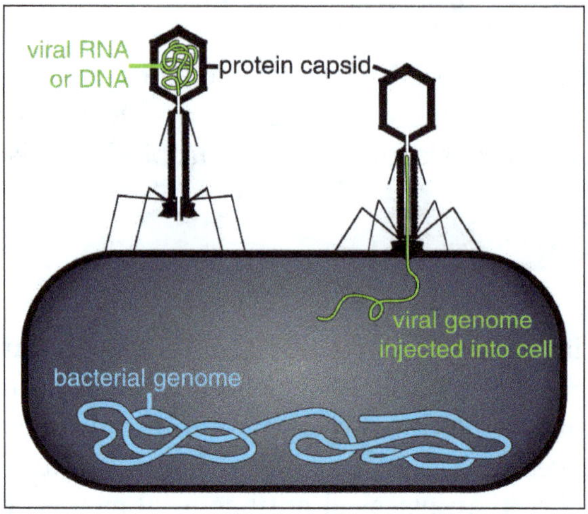

Phage injecting its genome into bacterial cell.

Phage therapy is another method for treating antibiotic-resistant strains of bacteria. Phage therapy infects pathogenic bacteria with their own viruses. Bacteriophages and their host ranges are extremely specific for certain bacteria, thus, unlike antibiotics, they do not disturb the host organism and intestinal microflora. Bacteriophages, also known simply as phages, infect and can kill bacteria and affect bacterial growth primarily during lytic cycles. Phages insert their DNA into the bacterium, where it is transcribed and used to make new phages, after which the cell will lyse, releasing new phage that are able to infect and destroy further bacteria of the same strain. The high specificity of phage protects "good" bacteria from destruction.

Some disadvantages to the use of bacteriophages also exist, however. Bacteriophages may harbour virulence factors or toxic genes in their genomes and, prior to use, it may be prudent to identify genes with similarity to known virulence factors or toxins by genomic sequencing. In addition, the oral and IV administration of phages for the eradication of bacterial infections poses a much higher safety risk than topical application. Also, there is the additional concern of uncertain immune responses to these large antigenic cocktails.

There are considerable regulatory hurdles that must be cleared for such therapies. Despite numerous challenges, the use of bacteriophages as a replacement for antimicrobial agents against MDR pathogens that no longer respond to conventional antibiotics, remains an attractive option.

Phytochemicals

Plants are an important source of antimicrobial compounds and traditional healers have long used plants to prevent or cure infectious diseases. There is a recent renewed interest into the use of natural products for the identification of new members of the 'antibiotic-ome' (defined as natural products with antibiotic activity), and their application in antibacterial drug discovery in the genomics era. Phytochemicals are the active biological component of plants and some phytochemicals including tannins, alkaloids, terpenoids, and flavonoids possess antimicrobial activity. Some antioxidant dietary supplements also contain phytochemicals (polyphenols), such as grape seed extract, and demonstrate *in vitro* anti-bacterial properties. Phytochemicals are able to inhibit peptidoglycan synthesis, damage microbial membrane structures, modify bacterial membrane surface hydrophobicity and also modulate quorum sensing. With increasing antibiotic resistance in recent years, the potential of new plant-derived antibiotics is under investigation.

New Antibiotics Development

Both the WHO and the Infectious Disease Society of America (IDSA) reported that the weak antibiotic pipeline does not match bacteria's increasing ability to develop resistance. The IDSA report noted that the number of new antibiotics approved for marketing per year had been declining and identified seven antibiotics against the Gram-negative bacilli (GNB) currently in phase 2 or phase 3 clinical trials. These drugs however, did not address the entire spectrum of resistance of GNB. According to the WHO fifty one new therapeutic entities (NTEs) - antibiotics (including combinations), are in phase 1-3 clinical trials as of May 2017. Recent entries in the clinical pipeline targeting multidrug-resistant Gram-positive pathogens has improved the treatment options due to marketing approval of new antibiotic classes, the oxazolidinones and cyclic lipopeptides. However, resistance to these antibiotics is certainly likely to occur, the need for the development new antibiotics against those pathogens still remains a high priority. Recent drugs in development that target Gram-negative bacteria have focused on re-working existing drugs to target specific microorganisms or specific types of resistance.

A few antibiotics have received marketing authorization in the last seven years. The cephalosporin ceftaroline and the lipoglycopeptides oritavancin and telavancin for the treatment of acute bacterial skin and skin structure infection and community-acquired bacterial pneumonia. The lipoglycopeptide dalbavancin and the oxazolidinone tedizolid has also been approved for use for the treatment of acute bacterial skin and skin structure infection. The first in a new class of narrow spectrum macrocyclic antibiotics, fidaxomicin, has been approved for the treatment of *C. difficile* colitis. New cephalosporin-lactamase inhibitor combinations also approved include ceftazidime-avibactam and ceftolozane-avibactam for complicated urinary tract infection and intra-abdominal infection.

- Ceftolozane/tazobactam (CXA-201; CXA-101/tazobactam): Antipseudomonal cephalosporin/β-lactamase inhibitor combination (cell wall synthesis inhibitor). FDA approved on 19 December 2014.

- Ceftazidime/avibactam (ceftazidime/NXL104): Antipseudomonal cephalosporin/β-lactamase inhibitor combination (cell wall synthesis inhibitor). In phase 3.
- Ceftaroline/avibactam (CPT-avibactam; ceftaroline/NXL104): Anti-MRSA cephalosporin/β-lactamase inhibitor combination (cell wall synthesis inhibitor).
- Imipenem/MK-7655: Carbapenem/β-lactamase inhibitor combination (cell wall synthesis inhibitor). In phase 2.
- Plazomicin (ACHN-490): Aminoglycoside (protein synthesis inhibitor). New Drug Application is under Priority Review by U.S. FDA.
- Eravacycline (TP-434): Synthetic tetracycline derivative/protein synthesis inhibitor targeting the ribosome. Development by Tetraphase, Phase 2 trials complete.
- Brilacidin (PMX-30063): Peptide defense protein mimetic (cell membrane disruption). In phase 2.

Streptomyces research is expected to provide new antibiotics, including treatment against MRSA and infections resistant to commonly used medication. Efforts of John Innes Centre and universities in the UK, supported by BBSRC, resulted in the creation of spin-out companies, for example Novacta Biosystems, which has designed the type-b lantibiotic-based compound NVB302 (in phase 1) to treat *Clostridium difficile* infections.

Possible improvements include clarification of clinical trial regulations by FDA. Furthermore, appropriate economic incentives could persuade pharmaceutical companies to invest in this endeavor. In the US, the Antibiotic Development to Advance Patient Treatment (ADAPT) Act was introduced with the aim of fast tracking the drug development of antibiotics to combat the growing threat of 'superbugs'. Under this Act, FDA can approve antibiotics and antifungals treating life-threatening infections based on smaller clinical trials. The CDC will monitor the use of antibiotics and the emerging resistance, and publish the data. The FDA antibiotics labeling process, 'Susceptibility Test Interpretive Criteria for Microbial Organisms' or 'breakpoints', will provide accurate data to healthcare professionals. According to Allan Coukell, senior director for health programs at The Pew Charitable Trusts, "By allowing drug developers to rely on smaller datasets, and clarifying FDA's authority to tolerate a higher level of uncertainty for these drugs when making a risk/benefit calculation, ADAPT would make the clinical trials more feasible."

Tetracycline

Tetracycline is used to treat infections caused by bacteria including pneumonia and other respiratory tract infections; ; certain infections of skin, eye, lymphatic, intestinal, genital and urinary systems; and certain other infections that are spread by ticks, lice, mites, and infected animals. It is also used along with other medications to treat acne. Tetracycline is also used to treat plague and tuleramia (serious infections that may be spread on purpose as part of a bioterror attack). It can also be used in patients who cannot be treated with penicillin to treat certain types of food poisoning, and anthrax (a serious infection that may be spread on purpose as part of a bioterror attack). Tetracycline is in a class of medications called tetracycline antibiotics. It works by preventing the growth and spread of bacteria.

Antibiotics such as tetracycline will not work for colds, flu, or other viral infections. Using antibiotics when they are not needed increases your risk of getting an infection later that resists antibiotic treatment.

How should this Medicine be used?

Tetracycline comes as a capsule to take by mouth. It is usually taken two or four times daily. Tetracycline should be taken on an empty stomach, at least 1 hour before or 2 hours after meals or snacks. Drink a full glass of water with each dose of tetracycline. Do not take tetracycline with food, especially dairy products such as milk, yogurt, cheese, and ice cream. Follow the directions on your prescription label carefully, and ask the doctor or pharmacist to explain any part you do not understand. Take tetracycline exactly as directed. Do not take more or less of it or take it more often than prescribed by the doctor.

Other uses for this Medicine

Tetracycline is also sometimes used to treat Lyme disease and malaria, and to prevent plague and tularemia in people who have been exposed to plague or tularemia germs. Talk to the doctor about the risks of using this medication for your condition. This medication is sometimes prescribed for other uses.

Precautions

Before taking tetracycline:

- Tell the and pharmacist if you are allergic to tetracycline, minocycline, doxycycline, demeclocycline, any other medications, or any of the ingredients in the tetracycline capsule. Ask your pharmacist for a list of the ingredients.
- Tell the and pharmacist what prescription and nonprescription medicines, vitamins, nutritional supplements, and herbal products you are taking or plan to take. Be sure to mention any of the following: anticoagulants ('blood thinners') such as warfarin (Coumadin, Jantoven), and penicillin.
- Be aware that antacids containing magnesium, aluminum, calcium, or sodium bicarbonate, calcium supplements, zinc products, iron products, and laxatives containing magnesium interfere with tetracycline, making it less effective. Take tetracycline 2 hours before or 6 hours after antacids, calcium supplements, zinc products, and laxatives containing magnesium. Take tetracycline 2 hours before or 4 hours after iron preparations and vitamin products that contain iron. Take tetracycline 2 hours before or after zinc containing products.

- Tell the doctor if you have or have ever had lupus (condition in which the immune system attacks many tissues and organs including the skin, joints, blood, and kidneys), or kidney disease.
- Tell the doctor if you are pregnant, plan to become pregnant, or are breastfeeding. If you become pregnant while taking tetracycline, call the doctor immediately. Tetracycline can harm the fetus.
- Plan to avoid unnecessary or prolonged exposure to sunlight and to wear protective clothing, sunglasses, and sunscreen. Tetracycline may make your skin sensitive to sunlight. Tell the doctor right away if you get a sunburn.
- You should know that when tetracycline is used during pregnancy or in babies or children up to age 8, it can cause the teeth to become permanently stained. Tetracycline should not be used in children under age 8 unless the doctor decides it is needed.

What Side Effects can this Medication Cause?

Tetracycline may cause side effects. Tell the doctor if any of these symptoms are severe or do not go away:

- Nausea.
- Vomiting.
- Diarrhea.
- Itching of the rectum or vagina.
- Swollen tongue.
- Black or hairy tongue.
- Sore or irritated throat.

Some side effects can be serious. If you experience any of these symptoms, call the doctor immediately:

- Headache.
- Blurred vision, seeing double, or loss of vision.
- Skin rash.
- Hives.
- Swelling of the face, throat, tongue, lips, and eyes.
- Difficulty breathing or swallowing.
- Joint stiffness or swelling.
- Unusual bleeding or bruising.

- Chest pain.
- A return of fever, sore throat, chills, or other signs of infection.
- Watery or bloody stools , stomach cramps, or fever during treatment or for up to two or more months after stopping treatment.

Tetracycline may cause other side effects. Call the doctor if you have any unusual problems while taking this medication.

What should I know about Storage and Disposal of this Medication?

Keep this medication in the container it came in, tightly closed, and out of reach of children. Store it at room temperature and away from light and excess heat and moisture (not in the bathroom).

Unneeded medications should be disposed of in special ways to ensure that pets, children, and other people cannot consume them. However, you should not flush this medication down the toilet. Instead, the best way to dispose of your medication is through a medicine take-back program. Talk to your pharmacist or contact your local garbage/ recycling department to learn about take-back programs in your community.

It is important to keep all medication out of sight and reach of children as many containers (such as weekly pill minders and those for eye drops, creams, patches, and inhalers) are not child-resistant and young children can open them easily. To protect young children from poisoning, always lock safety caps and immediately place the medication in a safe location – one that is up and away and out of their sight and reach.

References

- Heymann, D. L.; Aylward, R. B. (2006). Mass vaccination: When and why. Current Topics in Microbiology and Immunology. 304. Pp. 1–16. Doi:10.1007/3-540-36583-4_1. ISBN 978-3-540-29382-8. PMID 16989261
- "Tamiflu - Drugs.com". Www.drugs.com. Archived from the original on 21 December 2014. Re-trieved 16 January, 2017
- Immunization Safety Review Committee (2004).Immunization Safety Review: Vaccines and Autism. The National Academies Press. ISBN 0-309-09237-X
- Oseltamivir Phosphate". The American Society of Health-System Pharmacists. Archived from the original on 13 May 2016. Retrieved 8 January, 2017
- Wilson G, Block B (2004). Wilson and Gisvold's Textbook of Organic Medicinal and Pharmaceutical Chemistry. Philadelphia, Pa.: Lippincott Williams & Wilkins. ISBN 0-7817-3481-9
- Ameen M (March 2010). "Epidemiology of superficial fungal infections". Clinics in Dermatology. Elsevier Inc. 28 (2): 197–201. Doi:10.1016/j.clindermatol.2009.12.005. PMID 20347663
- Terbinafine Hydrochloride". The American Society of Health-System Pharmacists. Archived from the original on 21 December, 2016. Retrieved 8 December, 2016

3

Antimicrobial Therapies

Antimicrobial therapy helps in killing and inhibiting the growth of bacteria, fungi and protozoans. Empiric antimicrobial therapy and antimicrobial chemotherapy are the most important antimicrobial therapies. This chapter closely examines the key concepts related to antimicrobial therapy to provide an extensive understanding of the subject.

Antimicrobial agents are some of the most widely, and often injudiciously, used therapeutic drugs worldwide. Important considerations when prescribing antimicrobial therapy include obtaining an accurate diagnosis of infection; understanding the difference between empiric and definitive therapy; identifying opportunities to switch to narrow-spectrum, cost-effective oral agents for the shortest duration necessary; understanding drug characteristics that are peculiar to antimicrobial agents (such as pharmacodynamics and efficacy at the site of infection); accounting for host characteristics that influence antimicrobial activity; and in turn, recognizing the adverse effects of antimicrobial agents on the host. It is also important to understand the importance of antimicrobial stewardship, to know when to consult infectious disease specialists for guidance, and to be able to identify situations when antimicrobial therapy is not needed. By following these general principles, all practicing physicians should be able to use antimicrobial agents in a responsible manner that benefits both the individual patient and the community.

AST = antimicrobial susceptibility testing; CSF = Cerebrospinal fluid; ESBL = Extended-spectrum β-lactamase; G6PD = Glucose-6-phosphate dehydrogenase; HIV = human immunodeficiency virus; MIC = Minimum inhibitory concentration; MRSA = Methicillin-resistant Staphylococcus aureus; OPAT = Outpatient parenteral antimicrobial therapy; UTI = Urinary tract infection.

The terms antimicrobial, antibiotic, and anti-infective encompass a wide variety of pharmaceutical agents that include antibacterial, antifungal, antiviral, and antiparasitic drugs.

Selecting and Initiating an Antibiotic Regimen

Obtaining an Accurate Infectious Disease Diagnosis

An infectious disease diagnosis is reached by determining the site of infection, defining the host (eg, immunocompromised, diabetic, of advanced age), and establishing, when possible, a microbiological diagnosis. It is critical to isolate the specific pathogen in many serious, life-threatening infections, especially for situations that are likely to require prolonged therapy (eg, endocarditis, septic arthritis, disk space infection, and meningitis). Similarly, when a patient does not benefit from antimicrobial therapy chosen on the basis of clinical presentation, additional investigations are needed to determine the etiologic agent or exclude noninfectious diagnoses. To optimize an accurate microbiological diagnosis, clinicians should ensure that diagnostic specimens are properly obtained and promptly submitted to the microbiology laboratory, preferably before the institution of antimicrobial therapy. Infectious disease diagnoses also frequently rely on a detailed exposure history, as in the case of a patient with nonresolving pneumonia who has resided in or traveled to the southwestern United States where coccidioidomycosis is endemic. Although the microbiological diagnosis is ideally based on data such as bacterial or fungal culture or serologic testing, frequently the "most likely" microbiological etiology can be inferred from the clinical presentation. For example, cellulitis is most frequently assumed to be caused by streptococci or staphylococci, and antibacterial treatment can be administered in the absence of a positive culture. Similarly, community-acquired pneumonia that does not warrant hospitalization can also be treated empirically—with a macrolide or fluoroquinolone antibiotic—without performing specific diagnostic testing.2 Finally, noninfectious conditions should be considered in the differential diagnosis for infections, especially when the diagnosis is not clear-cut

Timing of Initiation of Antimicrobial Therapy

The timing of initial therapy should be guided by the urgency of the situation. In critically ill patients, such as those in septic shock, febrile neutropenic patients, and patients with bacterial meningitis, empiric therapy should be initiated immediately after or concurrently with collection of diagnostic specimens. In more stable clinical circumstances, antimicrobial therapy should be deliberately withheld until appropriate specimens have been collected and submitted to the microbiology laboratory. Important examples of this principle are subacute bacterial endocarditis and vertebral osteomyelitis/diskitis. Patients with these infections are frequently ill for a period of several days to weeks before presentation, and administration of antibiotic therapy should be delayed until multiple sets of blood cultures (in the case of endocarditis) or disk space aspirate and/or bone biopsy specimens (for osteomyelitis/diskitis) have been obtained. Premature initiation of antimicrobial therapy in these circumstances can suppress bacterial growth and preclude the opportunity to establish a microbiological diagnosis, which is critical in the management of these patients, who require several weeks to months of directed antimicrobial therapy to achieve cure.

Empiric vs. De initive Antimicrobial Therapy

Because microbiological results do not become available for 24 to 72 hours, initial therapy for infection is often empiric and guided by the clinical presentation. It has been shown that inadequate therapy for infections in critically ill, hospitalized patients is associated with poor outcomes, including greater morbidity and mortality as well as increased length of stay.[3,4] Therefore, a common approach is to use broad-spectrum antimicrobial agents as initial empiric therapy with the intent to cover multiple possible pathogens commonly associated with the specific clinical syndrome. This is true for both community- and hospital-acquired infections. For example, in an otherwise healthy young adult with suspected bacterial meningitis who is seen in the emergency department, the most likely pathogens would be Streptococcus pneumoniae and Neisseria meningitidis, and thus a combination of a third-generation cephalosporin (ceftriaxone) plus vancomycin would be recommended as empiric therapy.[5] Hospital-acquired infections are frequently related to the presence of invasive devices and procedures that result in loss of the normal barriers to infection, as is the case with intravascular catheter–associated bacteremia, ventilator-associated pneumonia, and catheter-associated urinary tract infections (UTIs). They are commonly caused by drug-resistant organisms, both gram-positive (eg, methicillin-resistant Staphylococcus aureus [MRSA]) and gram-negative (eg, Pseudomonas aeruginosa) bacteria, which are often endemic in hospitals because of the selection pressure from antimicrobial use. In selecting empiric antimicrobial therapy for such infections, clinicians should consider the following: (1) the site of infection and the organisms most likely to be colonizing that site (eg, intravascular catheter–associated bacteremia is frequently a result of colonization and infection caused by staphylococci present on the skin); (2) prior knowledge of bacteria known to colonize a given patient (eg, a screening nasal swab [currently conducted routinely by many hospitals before admitting patients to the intensive care unit] may indicate that the patient is colonized with MRSA); and (3) the local bacterial resistance patterns or antibiograms that are available for important pathogens at most hospitals.

Once microbiology results have helped to identify the etiologic pathogen and/or antimicrobial susceptibility data are available, every attempt should be made to narrow the antibiotic spectrum. This is a critically important component of antibiotic therapy because it can reduce cost and toxicity and prevent the emergence of antimicrobial resistance in the community. Antimicrobial agents with a narrower spectrum should be directed at the most likely pathogens for the duration of therapy for infections such as community-acquired pneumonia or cellulitis in the ambulatory setting because specific microbiological tests are not typically performed.

Interpretation of Antimicrobial Susceptibility Testing Results

When a pathogenic microorganism is identified in clinical cultures, the next step performed in most microbiology laboratories is antimicrobial susceptibility testing (AST). Antimicrobial susceptibility testing measures the ability of a specific organism to grow in the presence of a particular drug in vitro and is performed using guidelines established by the Clinical and Laboratory Standards Institute, a nonprofit global organization that develops laboratory process standards through extensive testing and clinical correlation. The goal of AST is to predict the clinical success or failure of the antibiotic being tested against a particular organism. Data are reported in the form of minimum inhibitory concentration (MIC), which is the lowest concentration of an antibiotic that inhibits visible growth of a microorganism, and are interpreted by the laboratory as "susceptible," "resistant," or "intermediate," according to Clinical and Laboratory Standards Institute criteria. A report of "susceptible" indicates that the isolate is likely to be inhibited by the usually achievable concentration of a particular antimicrobial agent when the recommended dosage is used for the particular site of infection. For this reason, MICs of different agents for a particular organism are not directly comparable. For example, MICs of 1 (susceptible) for ciprofloxacin and 2 (susceptible) for ceftriaxone against Escherichia coli do not imply that ciprofloxacin is twice as active as ceftriaxone. Instead, it indicates that concentrations achieved by giving recommended doses of both drugs are likely to be active against the organism. Although AST results are generally quite useful in narrowing the antibiotic regimen, AST has some limitations that should be kept in mind. First, it is important for both clinicians and laboratory personnel to be aware of the site of infection. For example, an isolate of S aureus could be reported as susceptible to cefazolin in vitro; however, if this particular isolate was obtained from the cerebrospinal fluid (CSF), cefazolin would not be an optimal therapeutic choice because it does not achieve therapeutic concentrations in the CSF. Clinical laboratories may provide different AST interpretations for different sites of infection (eg, meningitis and nonmeningitis AST results for S pneumoniae). In addition, some organisms carry enzymes that, when expressed in vivo, can inactivate antimicrobial agents to which the organism shows in vitro susceptibility. Although their presence is not immediately apparent from AST results, certain AST "patterns" can provide a clue to their existence. For example, extended-spectrum β-lactamases (ESBLs) in Enterobacteriaceae are enzymes that mediate resistance to almost all β-lactam agents except carbapenems (eg, meropenem or imipenem). Extended-spectrum β-lactamases can be difficult to detect because they have different levels of in vitro activity against various cephalosporins. In clinical practice, susceptibility to cephamycins (cefoxitin, cefotetan) but resistance to a third-generation cephalosporin (eg, cefpodoxime, cefotaxime, ceftriaxone, ceftazidime) or aztreonam should alert one to the possibility of ESBL production. The production of ESBL should also be suspected when treatment with β-lactams fails despite apparent in vitro susceptibility. This should lead to additional testing, which usually involves growing the bacteria in the presence of a third-generation cephalosporin alone and in combination with clavulanic acid (a β-lactamase inhibitor); enhanced bacterial inhibition with the addition of clavulanic acid indicates ESBL. When detected by the laboratory, these bacteria should be considered resistant to all β-lactam agents except the carbapenem class.

In general, it is good practice to communicate directly with the microbiology laboratory when antimicrobial susceptibility patterns appear unusual. It is also useful to be aware of the limitations of AST at the local laboratory, particularly in smaller hospitals (eg, testing of relatively newer agents [such as daptomycin for gram-positive cocci] might not be routinely performed or reported but could be available on request).

Bactericidal vs. Bacteriostatic Therapy

A commonly used distinction among antibacterial agents is that of bactericidal vs bacteriostatic agents. Bactericidal drugs, which cause death and disruption of the bacterial cell, include drugs that primarily act on the cell wall (eg, β-lactams), cell membrane (eg, daptomycin), or bacterial DNA (eg, fluoroquinolones). Bacteriostatic agents inhibit bacterial replication without killing the organism. Most bacteriostatic drugs, including sulfonamides, tetracyclines, and macrolides, act by inhibiting protein synthesis. The distinction is not absolute, and some agents that are bactericidal against certain organisms may only be bacteriostatic against others and vice versa. In most cases, this distinction is not significant in vivo; however, bactericidal agents are preferred in the case of serious infections such as endocarditis and meningitis to achieve rapid cure.

Use of Antimicrobial Combinations

Although single-agent antimicrobial therapy is generally preferred, a combination of 2 or more antimicrobial agents is recommended in a few scenarios.

When agents exhibit synergistic activity against a microorganism. Synergy between antimicrobial agents means that, when studied in vitro, the combined effect of the agents is greater than the sum of their independent activities when measured separately.8 For example, the combination of certain β-lactams and aminoglycosides exhibits synergistic activity against a variety of gram-positive and gram-negative bacteria9 and is used in the treatment of serious infections, for which rapid killing is essential (eg, treatment of endocarditis caused by Enterococcus species with a combination of penicillin and gentamicin). In this setting, the addition of gentamicin to penicillin has been shown to be bactericidal, whereas penicillin alone is only bacteriostatic and gentamicin alone has no significant activity. For certain streptococci, similar synergistic combinations that result in more rapid clearance of the infecting microorganism can also be used to shorten the course of antimicrobial therapy (eg, for endocarditis due to viridans group streptococci, a combination of penicillin or ceftriaxone with gentamicin for 2 weeks can be as effective as penicillin or ceftriaxone alone for 4 weeks).

When critically ill patients require empiric therapy before microbiological etiology and/or antimicrobial susceptibility can be determined. Antibiotic combinations are used in empiric therapy for health care–associated infections that are frequently caused by bacteria resistant to multiple antibiotics. Combination therapy is used in this setting to ensure that at least 1 of the administered antimicrobial agents will be active against the suspected organism(s). For example, when a patient who has been hospitalized for several weeks develops septic shock and blood cultures are reported to be growing gram-negative bacilli, it would be appropriate to provide initial therapy with 2 agents that have activity against gram-negative bacilli, particularly P aeruginosa, which is both a common nosocomial pathogen and frequently resistant to multiple agents—in this case, a combination of an antipseudomonal β-lactam with a fluoroquinolone or aminoglycoside could be used.

To extend the antimicrobial spectrum beyond that achieved by use of a single agent for treatment of polymicrobial infections. When infections are thought to be caused by more than one organism, a combination regimen may be preferred because it would extend the antimicrobial spectrum beyond that achieved by a single agent. For example, most intra-abdominal infections are usually caused by multiple organisms with a variety of gram-positive cocci, gram-negative bacilli, and anaerobes. Antimicrobial combinations, such as a third-generation cephalosporin or a fluoroquinolone plus metronidazole, can be used as a potential treatment option in these cases and can sometimes be more cost-effective than a comparable single agent (eg, a carbapenem).

To prevent emergence of resistance. The emergence of resistant mutants in a bacterial population is generally the result of selective pressure from antimicrobial therapy. Provided that the mechanisms of resistance to 2 antimicrobial agents are different, the chance of a mutant strain being resistant to both antimicrobial agents is much lower than the chance of it being resistant to either one. In other words, use of combination therapy would provide a better chance that at least one drug will be effective, thereby preventing the resistant mutant population from emerging as the dominant strain and causing therapeutic failure. This is why combination drug therapy is used as the standard for treatment of infections such as tuberculosis and the human immunodeficiency virus (HIV) when treatment duration is likely to be prolonged, resistance can emerge relatively easily, and therapeutic agents are limited.

Host Factors to be Considered in Selection of Antimicrobial Agents

Although it is helpful for clinicians to gain familiarity with a few specific antimicrobial agents, a "one size fits all" approach is not appropriate in antimicrobial selection, and several host factors must be taken into account. Published guidelines on appropriate dose adjustments for individual antimicrobial agents are available from a variety of sources.

Renal and Hepatic Function: Because the kidney and the liver are the primary organs responsible for elimination of drugs from the body, it is important to determine how well they are functioning during antimicrobial administration. In most cases, one is concerned with dose reduction to prevent accumulation and toxicity in patients with reduced renal or hepatic function. However, sometimes doses might need to be increased to avoid underdosing young healthy patients with rapid renal elimination or those with rapid hepatic metabolism due to enzyme induction by concomitant use of drugs such as rifampin or phenytoin.

Age: Patients at both extremes of age handle drugs differently, primarily due to differences in body size and kidney function. Most pediatric drug dosing is guided by weight. In geriatric patients, the serum creatinine level alone is not completely reflective of kidney function, and the creatinine clearance should be estimated by factoring in age and weight for these patients.

Genetic Variation: Genetic susceptibility to the adverse effects of antimicrobial agents, which has been demonstrated for several antimicrobial agents, is occasionally significant enough to warrant testing for such variability before administration of certain drugs.8 For example, the antiretroviral drug abacavir, which has become part of the standard combination treatment for HIV infection, is associated with a well-described and potentially fatal hypersensitivity reaction that can manifest with any combination of fever, rash, abdominal pain, and respiratory distress. The risk of experiencing this reaction has been shown to be significantly higher in patients with the human leukocyte antigen allele HLA-B*5701,14 and current HIV treatment guidelines recommend routine screening for the presence of this genetic susceptibility in patients before prescribing this drug. Another example is that of glucose-6-phosphate dehydrogenase (G6PD) deficiency, which can result in hemolysis in individuals when exposed to certain antimicrobial agents, such as dapsone, primaquine, and nitrofurantoin. These drugs should be avoided in those known to be deficient in G6PD, and it is advisable to test for this predisposition in patients who might have a higher risk of G6PD deficiency (eg, African Americans) before prescribing these agents. Many antimicrobial agents are handled by the hepatic cytochrome P450 system, and although variation in expression of these enzymes occurs, insufficient data are available to recommend routine clinical testing to guide antimicrobial dosing.

Pregnancy and Lactation: Special considerations for the use of antimicrobial agents in pregnancy relate to both the mother and the fetus. In the case of the mother, increases in plasma volume and renal blood flow, especially by the third trimester, can result in more rapid clearance and lower serum levels of pharmaceutical agents, including antimicrobial agents. However, data to support the clinical relevance of this change are sparse, and higher antimicrobial doses are not routinely recommended in the third trimester of pregnancy. Some experts recommend an increased dose of several protease inhibitors for the management of HIV infection in pregnancy. In the case of the developing fetus, many antimicrobial agents can be either teratogenic or otherwise toxic to the fetus. Penicillins, cephalosporins, and macrolides have historically been the most commonly used antimicrobial agents considered safe in pregnancy, and a recent multicenter study of more than 13,000 women with pregnancies affected by birth defects found no association between adverse outcomes and these particular antimicrobial agents. Other drugs, such as tetracyclines and chloramphenicol, have well-described fetal or neonatal adverse effects and should be avoided. In general, however, human studies on the safety of many antimicrobial agents in pregnancy and lactation are limited, and antimicrobial agents should be prescribed with caution.

History of Allergy or Intolerance: A history of antimicrobial allergy or intolerance should be routinely obtained in the evaluation and management of infection.

History of Recent Antimicrobial Use: Eliciting a history of exposure to antimicrobial agents in the recent past (approximately 3 months) can also help in selection of antimicrobial therapy. Because the causative microorganism for a current episode of infection emerged under the selective pressure of a recently used antimicrobial agent, it is likely to be resistant to that drug and/or drug class, and an alternative agent should be used.

Oral vs. Intravenous Therapy

Patients hospitalized with infections are often treated with intravenous antimicrobial therapy because their admission is often prompted by the severity of their infection. However, patients with mild to moderate infections who require hospitalization for other reasons (eg, dehydration, pain control, cardiac arrhythmias) and have normal gastrointestinal function are candidates for treatment with well-absorbed oral antimicrobial agents (eg, treatment of pyelonephritis and community-acquired pneumonia with oral fluoroquinolones). Furthermore, patients initially treated with parenteral therapy can be safely switched to oral antibiotics when they become clinically stable. When using oral therapy for invasive infections (such as pneumonia, pyelonephritis, or abscesses), clinicians are advised to select an agent that has excellent absorption and bioavailability (ie, the percentage of the oral dose that is available unchanged in the serum). Examples of antibiotics with excellent bioavailability are fluoroquinolones, linezolid, trimethoprim-sulfamethoxazole, and metronidazole. For more serious infections, such as infective endocarditis and central nervous system infections (eg, meningitis), in which high serum or CSF drug concentrations are desired, a switch to oral therapy is less reliable and not generally recommended.

Pharmacodynamic Characteristics

Along with host factors, the pharmacodynamic properties of antimicrobial agents may also be important in establishing a dosing regimen. Specifically, this relates to the concept of time-dependent vs concentration-dependent killing. Drugs that exhibit time-dependent activity (β-lactams and vancomycin) have relatively slow bactericidal action; therefore, it is important that the serum concentration exceeds the MIC for the duration of the dosing interval, either via continuous infusion or frequent dosing. In contrast, drugs that exhibit concentration-dependent killing (aminoglycosides, fluoroquinolones, metronidazole, and daptomycin) have enhanced bactericidal activity as the serum concentration is increased. With these agents, the "peak" serum concentration, and not the frequency of the dosing interval, is more closely associated with efficacy. To illustrate the impact of this distinction on dosing options, we can take the example of a 70-year-old woman with a creatinine clearance estimated to be 30 mL/min who is being treated with ciprofloxacin for pyelonephritis caused by E coli. Antimicrobial dosing guidelines suggest that a dose of either 250 mg orally every 12 hours or 500 mg every 24 hours is an acceptable modification for her reduced kidney function. However, given that ciprofloxacin exhibits concentration-dependent killing, selection of the latter dosing schedule would be more appropriate. In contrast, if the same patient were being treated with intravenous ampicillin, for which the time above the MIC is more closely related to efficacy, a dose of 1 g every 4 hours would be preferable to 2 g every 8 hours.

Efficacy at the Site of Infection

In addition to possessing in vitro antimicrobial activity and achieving adequate serum levels, the efficacy of antimicrobial agents depends on their capacity to achieve a concentration equal to or greater than the MIC at the site of infection and modification of activity at certain sites. Antimicrobial concentrations attained at some sites (eg, ocular fluid, CSF, abscess cavity, prostate, and bone) are often much lower than serum levels. For example, first- and second-generation cephalosporins and macrolides do not cross the blood-brain barrier and are not recommended for central nervous system infections. Fluoroquinolones achieve high concentrations in the prostate and are preferred oral agents for the treatment of prostatitis. Daptomycin, an excellent bactericidal agent against gram-positive bacteria, is not useful for treatment of pneumonia (eg, pneumococcal pneumonia) because it is inactivated by lung surfactant. Many antibiotics (eg, aminoglycosides) are less active in the low-oxygen, low-pH, and high-protein environment of abscesses, and drainage of abscesses to enhance antimicrobial efficacy is recommended when possible. Agents in the same class can differ from one another; for example, moxifloxacin does not achieve significant urinary concentrations because of its low renal excretion and is therefore not suitable for treatment of UTIs; in contrast, both levofloxacin and ciprofloxacin are excellent choices for UTIs caused by susceptible bacteria. The presence of foreign bodies at the site of infection also affects antimicrobial activity.

Selection of Antimicrobial Agents for Outpatient Parenteral Antimicrobial Therapy

To decrease cost, and with the help of advances both in antimicrobial agents and in technology to assist antimicrobial administration, prolonged treatment of serious infections with intravenous or parenteral antimicrobial agents has increasingly shifted away from the hospital to the outpatient setting, and guidelines to assist with delivery of high-quality outpatient parenteral antimicrobial therapy (OPAT) have been developed. Therapy can be provided via one of several types of indwelling central venous access catheters (a peripherally inserted central catheter is most frequently used) and can be delivered at an infusion center, by a home-visiting nurse, by self-administration, or in a nursing home. In addition to the general principles for selection of antimicrobial agents, OPAT requires some further considerations. First, other things being equal, an agent that requires less frequent administration is preferred. For example, for the treatment of osteomyelitis or other serious infections caused by methicillin- or oxacillin-sensitive S aureus, cefazolin is frequently used in favor of nafcillin or oxacillin because it allows administration every 8 hours. Its use makes treatment outside the hospital setting much more feasible than the administration every 4 hours required for the other drugs. Agents with once- or twice-daily dosing have gained popularity for OPAT and include ceftriaxone, ertapenem, vancomycin, and daptomycin. An alternative for most β-lactams, which require frequent dosing, is use of a continuous infusion pump; however, such a device can frequently be cost-prohibitive. Second, the agent must possess chemical stability and should last for about 24 hours after mixing to allow enough time for delivery and administration. As an important illustration of the principle, the use of intravenous ampicillin for OPAT via self-administration or continuous infusion is often precluded because of a short (approximately 8-hour) stability period at room temperature. Ampicillin or penicillin (in combination with an aminoglycoside) is the drug of choice for endocarditis caused by penicillin-sensitive enterococci; therefore, OPAT for this type of infection usually necessitates either nursing home stay or investment in a continuous infusion device (for penicillin only). Third, agents with minimal toxicity or predictable toxicity amenable to monitoring are preferred as OPAT is generally used in the context of longer-term antimicrobial therapy. Finally, when possible, provided adherence can be expected, consideration should be given to using oral agents in the outpatient setting.

Use of Therapeutic Drug Monitoring

Monitoring serum concentrations for drugs is most useful for medications that have a fairly narrow therapeutic index, which is the ratio of the toxic to the therapeutic dose. Fortunately, most antimicrobial agents have a wide therapeutic index, allowing standard doses to be used, with predictable modifications on the basis of age, weight, and renal and hepatic function. However, certain antimicrobial agents require monitoring of serum levels because the therapeutic window is narrow. This could be due primarily to toxicity at high levels (eg, aminoglycosides) or therapeutic failure at low drug levels (eg, vancomycin) but is usually a combination of both (eg, voriconazole). In some cases, the use of serum drug level monitoring is supported by its beneficial effect on clinical outcomes (eg, voriconazole in the treatment of invasive fungal infections).

Considerations for Continuing Antibiotic Therapy

Duration of Antimicrobial Therapy

The duration of therapy for many infections has long been based on anecdotal data and expert opinion. In view of the deleterious effects of prolonged courses of antimicrobial agents, including the potential for adverse reactions, problems with adherence, selection of antibiotic-resistant organisms, and high cost, a number of studies have tried to define the optimal duration of therapy, with an emphasis on shorter courses of therapy. In other situations, a longer duration of therapy is clearly warranted (eg, 4-6 weeks for endocarditis, osteomyelitis, and intra-abdominal abscesses, and weeks to months for invasive fungal infections) to achieve cure and prevent relapse. In many such infections, treatment duration has to be carefully individualized on the basis of clinical and radiologic response and may require the guidance of an expert in infectious diseases.

Assessment of Response to Treatment

Response to treatment of an infection can be assessed using both clinical and microbiological parameters. Clinical parameters of improvement include symptoms and signs (eg, a decrease in fever, tachycardia, or confusion), laboratory values (eg, decreasing leukocyte count), and radiologic findings (eg, decrease in the size of an abscess). Although radiologic criteria are commonly used in assessing response to infectious disease therapy, radiologic improvement can frequently lag behind clinical improvement, and routine radiographic follow-up of all infections is not always necessary.

Bacteremia is the most common scenario in which microbiological response is closely assessed because clearance of the bloodstream is as important as clinical improvement. Persistent bacteremia can often be the only clue to the presence of an inadequately treated source or to the existence or development of endovascular infection (such as endocarditis or an intravascular device infection). Persistent bacteremia can also be associated with the emergence of antimicrobial resistance and should always be investigated.

Adverse Effects

Although the term antimicrobial allergy is frequently used synonymously with adverse reaction or adverse effect, allergic reactions constitute only one subset of adverse reactions to antimicrobial agents.

Table: Classification of the adverse effects of antimicrobial drugs.

Direct
Allergy
Toxicity
Drug-drug interaction
Indirect
Effect on commensal flora
Human } Clostridium Animal } Increased chance of infection with drug-resistant pathogens
Effects on environmental flora

Allergic or hypersensitivity reactions can be either immediate (IgE-mediated) or delayed and usually manifest as a rash; anaphylaxis is the most severe manifestation of IgE-mediated allergy. In a recent national study of the prevalence of adverse drug effects, antibiotics were implicated in 19% of all emergency department visits for drug-related adverse events, and 79% of all antibiotic-associated adverse events were classified as allergic reactions. Although a history of serious allergic reaction should be carefully documented to avoid inadvertent administration of the same drug or another drug in the same class, self-report of antibiotic allergies can be quite unreliable—it has been shown that only 10% to 20% of patients reporting a history of penicillin allergy were truly allergic when assessed by skin testing. Historical details should be elicited to help distinguish allergic from nonallergic reactions and IgE-mediated from delayed reactions because failure to do so can result in unnecessary avoidance of the most effective, narrow-spectrum, and cost-effective antimicrobial agent (eg, use of vancomycin in place of a β-lactam). Although no single test or clinical finding leads to a diagnosis of antibiotic allergy, a negative skin test can reliably exclude the possibility of developing an IgE-mediated reaction (such as anaphylaxis) and help optimize antibiotic use.Both clinicians and patients should understand that a negative skin test does not mean that a patient is not at risk for developing a non–IgE-mediated delayed allergic reaction, but that in many circumstances the benefit of receiving a more appropriate antibiotic would outweigh the risk of a less significant allergic reaction. If an ongoing reaction is attributed to an antimicrobial drug allergy, this usually requires discontinuation of the offending agent. Related drugs (eg, cephalosporins in patients with a history of penicillin allergy) can be used under careful observation, provided that the reaction is not severe or the skin test is negative. In some cases, if the offending agent is the only or highly preferred agent, desensitization may be necessary. Desensitization involves administration of the drug in progressively increasing doses given by mouth; protocols are available for certain agents, such as β-lactams and sulfonamides, and should be guided by experts in allergic diseases.

Nonallergic drug toxicity is usually, but not always, associated with higher doses and/or prolonged use and is particularly noted in patients with poor kidney or liver function that results in impaired clearance. Examples include nephrotoxicity with aminoglycosides, neurotoxicity of penicillins, and peripheral neuropathy with prolonged use of metronidazole; these potential adverse effects need to be discussed with patients before initiation of therapy. For patients receiving prolonged systemic antimicrobial therapy, periodic clinical and laboratory monitoring is also recommended, particularly for those drugs that cause predictable toxicity with increasing duration of use (eg, monitoring complete blood cell count, including white blood cell differential, with β-lactams, trimethoprim-sulfamethoxazole, and linezolid; creatine kinase level with daptomycin; and creatinine level with aminoglycoside and β-lactams). In addition, drug doses should be adjusted in response to changes in creatinine level to avoid toxicity and attain optimal serum concentrations.

Many antimicrobial agents interact with other drugs to increase or decrease their serum levels and effects. This is frequently the case with antimicrobial agents that are metabolized by and/or affect the cytochrome P450 enzyme system (eg, rifampin is a powerful inducer, whereas macrolides and azole antifungal agents are inhibitors of cytochrome P450 enzymes). Clinicians should always remain alert to the possibility of such interactions of antimicrobial agents with other drugs, and it is advisable to review a patient's medication list when prescribing antimicrobial agents. Certain drug combinations can also cause additive toxicity, as exemplified by the concomitant use of amphotericin and gentamicin, which can significantly increase the risk of nephrotoxicity.

Special Situations in Infectious Disease Therapy

Antimicrobial Therapy for Foreign Body–associated Infections

Prosthetic implants and devices are increasingly being used in modern medical treatment. An unfortunate consequence of this increased use is the emergence of infections associated with the placement of such devices, involving both temporary (eg, urinary catheter, central venous catheter) and permanent (eg, prosthetic joint, artificial heart valve) implants. One of the important characteristics of device-related infection is the formation of biofilms, which have been described as "a structured community of bacterial cells enclosed in a self-produced polymeric matrix and adherent to an inert or living surface." Bacteria growing in biofilms have been shown to be relatively protected from the effects of antimicrobial therapy, probably as a result of alteration of their metabolic state. Primary care physicians should be aware of this because prolonged antibiotic treatment for these infections can be ineffective, associated with adverse effects, and result in the emergence of resistant strains of organisms. Certain agents (eg, rifampin and fluoroquinolones) have better activity against staphylococci in biofilms and are recommended in the management of infections of prosthetic valves and joints caused by these organisms. However, because of the difficulty of eradicating infections with antimicrobial therapy alone, removal of the implant is often necessary for cure. As an alternative, for patients unable to tolerate implant removal, long-term suppressive antimicrobial therapy is sometimes used, with variable success. It is advisable to involve an infectious diseases expert in the management of infections associated with implanted foreign bodies.

Use of Antimicrobial Agents as Prophylactic or Suppressive Therapy

In an ideal scenario for use of an antimicrobial agent as prophylactic treatment, the infection would occur predictably in a certain setting and would be well known to be associated with a specific organism or organisms, and an effective antimicrobial agent would be available with no or limited long-term toxicity and with little likelihood of leading to the emergence of resistance. Not surprisingly, such scenarios are relatively rare. However, antimicrobial prophylaxis is appropriate in some instances, a discussion of which follows.

Presurgical antimicrobial prophylaxis: Antimicrobial prophylaxis is used to reduce the incidence of postoperative surgical site infections. Patients undergoing procedures associated with high infection rates, those involving implantation of prosthetic material, and those in which the consequences of infection are serious should receive perioperative antibiotics. The antibiotic(s) should cover the most likely organisms and be present in the tissues when the initial incision is made, and adequate serum concentrations should be maintained during the procedure. A single dose of a cephalosporin (such as cefazolin) administered within 1 hour before the initial incision is appropriate for most surgical procedures; this practice targets the most likely organisms (ie, skin flora), while avoiding unnecessary broad-spectrum antimicrobial therapy. Duration of prophylaxis for surgical site infection should not exceed 24 hours in most cases.

Antimicrobial prophylaxis in immunocompromised patients: Immunocompromised patients, particularly those with HIV infection/AIDS, those who are undergoing chemotherapy for cancer, or those who are receiving immunosuppressive therapy after organ transplant, are at increased risk of infection. These infections are caused by predictable organisms at an increased frequency and/or associated with high mortality (eg, invasive aspergillosis associated with prolonged neutropenia, Pneumocystis pneumonia in the setting of impaired cell-mediated immunity. In these specific settings, evidence supports the use of prolonged antimicrobial prophylaxis until immune markers are restored (eg, trimethoprim-sulfamethoxazole to prevent Pneumocystis pneumonia).

Antimicrobial prophylaxis to prevent transmission of communicable pathogens to susceptible contacts: Antimicrobial agents can be prescribed prophylactically to prevent transmission of pathogens to susceptible contacts; for example, antiviral agents can be used to limit the spread of influenza in nursing home residents, ciprofloxacin can be given to close contacts of a patient with meningitis caused by N meningitidis, and macrolides can be prescribed to reduce transmission of pertussis.

Antimicrobial prophylaxis before dental and other invasive procedures in patients susceptible to bacterial endocarditis: It should be noted that guidelines recommending antimicrobial prophylaxis in this setting have recently been updated and limit such use to only a few very high-risk scenarios—prosthetic valves, prior endocarditis, or congenital heart disease before surgical correction.

Traumatic injuries with a high probability of infectious complications: Certain types of injuries pose a particularly high risk of infection because of disruption of normal barriers and/or delivery of a high inoculum of pathogenic organisms (eg, antibiotic prophylaxis has been shown to be of some benefit and is recommended for certain types of animal bites and after penetrating brain injury). An example of inappropriate antimicrobial "prophylaxis" is prolonged antimicrobial use in those with open but not infected wounds, including surgical wounds. No consensus has yet been reached on the use of antimicrobial prophylaxis in some other settings, such as before invasive procedures in patients with prosthetic joints.

Judicious use of Antimicrobial Agents

Cost Considerations in Antibiotic Selection and Antimicrobial Stewardship

The "cost" of an antimicrobial agent is dependent on many factors in addition to the purchase price of a particular agent and may include administration costs, prolonged hospitalization as a consequence of adverse effects, the cost of serum concentration monitoring, and clinical efficacy. One strategy that can significantly reduce cost is the switch from intravenous to oral therapy. Oral therapy is generally less expensive, potentially associated with fewer adverse effects, and can result in considerable cost savings by facilitating earlier dismissal and a shortened hospital stay.Even if the purchase price of an oral agent is greater than its parenteral equivalent, the reduction in hospital stay can result in significant cost savings. This has been demonstrated for oral linezolid when compared with intravenous vancomycin for the treatment of complicated skin and soft tissue infections caused by MRSA.

Cost considerations in the selection and continuation of appropriate antimicrobial therapy in acute care hospitals are part of a broader activity that is referred to as antimicrobial stewardship. Antimicrobial stewardship programs are aimed at "optimizing antimicrobial selection, dosing, route, and duration of therapy to maximize clinical cure or prevention of infection while limiting the unintended consequences, such as the emergence of resistance, adverse drug events, and cost." These programs are usually coordinated by a team of infectious disease physician(s) and pharmacist(s) and are often computer-based. Some components recommended for these programs include the following: prospective audit and feedback of antimicrobial prescriptions to clinicians, formulary restriction, education, use of clinical order sets and guidelines, de-escalation of therapy, and intravenous to oral antimicrobial conversion when appropriate.Clinicians should make it a priority to become aware of such programs in their institutions.

Preventing Emergence of Antibiotic Resistance

The widespread—and often inappropriate—use of antimicrobial agents is the single most important cause of the emergence of drug resistance, both in the community and hospital settings. Prior antibiotic exposure has been shown to be the most frequent risk factor for the development of community-acquired respiratory infections caused by drug-resistant S pneumoniae. This is not surprising because acute upper respiratory illnesses account for the highest proportion of ambulatory antibiotic prescriptions, with most being dispensed in situations in which antibiotics were not even indicated. Clearly, the emergence of antimicrobial resistance can be prevented or delayed through judicious prescribing, which can be characterized as follows: avoidance of antibiotic treatment for community-acquired, mostly viral, upper respiratory tract infections; use of narrow-spectrum antibiotics when possible; and use of antibiotics for the shortest duration that is effective for the treatment of a particular clinical syndrome. In the past few years, interest has been increasing in the development of rapid and accurate diagnostic tests for detection of viral respiratory pathogens with the ability to distinguish between viral and bacterial infections, such as measurement of procalcitonin levels and nucleic acid tests. Although not yet widely available in clinical practice, these tests have the potential to curtail the use of antibacterial agents for clinical syndromes that are clearly caused by viruses.

Common Misuses of Antibiotics

Prolonged empiric antimicrobial treatment without clear evidence of infection: One of the most common mistakes in antimicrobial use is continuing to add or switch antibiotics when a patient does not appear to be responding to therapy, even though there is no clear evidence of an infectious disease. Many noninfectious, inflammatory, or neoplastic syndromes can present with symptoms and signs that mimic infectious diseases. Examples include adult-onset Still disease and other connective tissue disorders that can present with high fever; drug-induced fever; the fever associated with pulmonary embolism; lymphoma; and Wegener granulomatosis, which can present with fever, cavitary pulmonary nodules, and recurrent sinusitis.

Treatment of a positive clinical culture in the absence of disease: Colonization with potentially pathogenic organisms without any associated manifestation of disease occurs frequently in certain populations (eg, colonization of the urinary tract in women of advanced age or in the presence of an indwelling urinary catheter, colonization of endotracheal tubes in mechanically ventilated patients, and colonization of chronic wounds). Appropriate management in these situations involves obtaining cultures from these sites only when indicated and avoiding treatment of a "positive" culture result when symptoms and signs of active infection are absent (eg, asymptomatic bacteriuria).

Failure to narrow antimicrobial therapy when a causative organism is identified: Initial therapy is often empiric and relies on broad-spectrum agents until culture or other tests help determine the microbiological etiology. Once culture and susceptibility data are available, an antibiotic with the narrowest possible spectrum should be selected for continuation of therapy. Often, however, this does not occur, particularly if the patient has improved while receiving empiric therapy, and the physician is uncomfortable about changing therapy in the face of clinical improvement.

Prolonged prophylactic therapy: Infection can be prevented in certain situations by the prophylactic use of antimicrobial agents (eg, presurgical prophylaxis). However, in most cases, guidelines support the use of a single, preoperative dose of an antimicrobial agent. Prolonged "prophylaxis" simply sets the stage for the emergence of antimicrobial resistance. For example, the common practice of prolonging antimicrobial therapy until the removal of surgical drains is not evidence based.

Excessive use of certain antimicrobial agents: The frequent use of certain agents (or classes of antimicrobial agents) in a hospital or other health care setting can result in selection of organisms that are resistant to that particular antibiotic. For example, the increased use of fluoroquinolones during the past decade is thought to be, in part, responsible for the epidemic of a fluoroquinolone-resistant strain of C difficile, the most common cause of nosocomial infectious diarrhea. More recently, an increase in levofloxacin use as initial therapy for UTI as a result of policy change at a single institution was found to have led to a rapid increase in fluoroquinolone resistance among outpatient urinary E coli isolates at that institution. For this reason, those involved in antimicrobial stewardship should avoid the excessive prescribing of a single class of antibiotic.

Goals of Antimicrobial Therapy

Infectious disease remains a major cause of death and morbidity. According to data from the World Health Organization (WHO) in 1998, infection was the second most common cause of death, causing 25% of deaths worldwide. However, in developing countries, infectious diseases accounted for almost half of all deaths. Importantly, infection was responsible for more than 60% of all deaths among children aged 0-4 years. The WHO data also show that almost 90% of deaths from infectious diseases were caused by six major diseases, of which acute respiratory tract infection was the main contributor.

Although the development of antimicrobials has greatly reduced mortality and morbidity from infectious diseases, at least in the developed world, the emergence of resistance to antimicrobials threatens to undermine these advances. In particular, the global emergence of bacterial resistance in respiratory tract infections is of considerable concern. All of the major bacterial pathogens involved in these infections, Streptococcus pneumoniae, Haemophilus injluenzae, and Movaxella cataruhalis, have developed resistance to one or more antimicrobial classes. For example, S. pneumoniae has developed resistance to all classes of antimicrobial agent used to treat respiratory tract infections, although resistance prevalences vary between regions and for different antimicrobial agents.

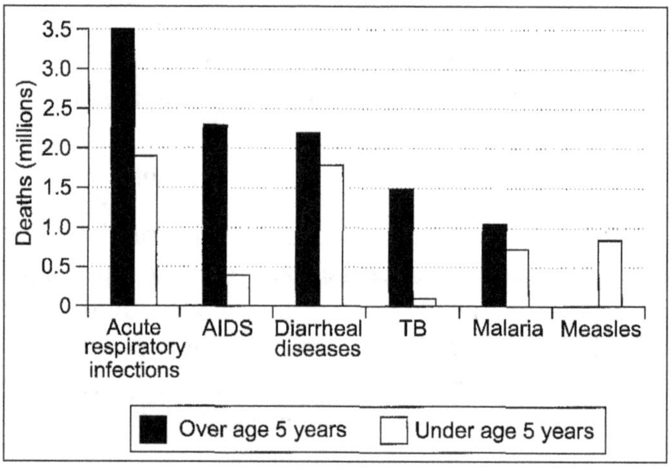

Leading causes of mortality due to infectious diseases.

The history of antimicrobial use in respiratory tract infection and the emergence of resistance indicates that there are still serious problems in defining and choosing the best (most appropriate) antimicrobials for clinical therapy. In order to achieve successful therapy of respiratory tract infection, we need to define clear goals for 'success', and an evidence-based approach for determining the potential of different therapies to achieve these goals. This supplement outlines the importance of bacterial eradication as the primary goal of antimicrobial therapy. Pharmacokinetic (PK) and pharmacodynamic (PD) analyses can be used to predict bacterial eradication and applied to the assessment and choice of agents currently used for the treatment of respiratory tract infections as well as the development of improved formulations and new antimicrobials. Finally, the need for changes in prescribing practice, driven by PK/PD predictions of bacterial eradication to maximize clinical outcomes and minimize the emergence and spread of resistance, is examined.

Appropriate Prescribing and Drug Development

The continued development and spread of antimicrobialresistant bacterial strains creates an urgent need for the development of new antimicrobials and the appropriate use of currently available agents. Few new antimicrobials are now reaching the market;

development timelines are lengthy, often more than a decade, and new agents are untried in clinical practice. There are many examples showing that the safety profile of a new agent cannot be guaranteed based on the profiles of other antimicrobials in the same class. In addition, for new agents developed from existing antimicrobial classes, cross-resistance with other members of the class may also be a problem. The continued development of currently available agents through the introduction of new formulations and doses is attractive, as it is likely to be much faster than developing new antimicrobials, and has the advantage of a known efficacy and safety record. However, new formulations/doses must be able to overcome existing bacterial resistance.

Limitations of Clinical Assessment of Antimicrobial Efficacy

Before the use of current antimicrobials can be optimized, a satisfactory way to assess their efficacy must be found. Antimicrobials are usually assessed formally in clinical trials, and prescribing recommendations are often made based on these trials. Although clinical trials constitute the accepted way of proving the therapeutic efficacy of antimicrobials, their limitations must always be borne in mind, in order to avoid reaching unsupported conclusions.

Study design and sample size are problems in many clinical trials. Trials of new agents in comparison to the established therapy are often designed to show nonnferiority only, rather than superiority. This is because more than 700 patients are needed in each of the two arms of the trial in order to show that one agent has a clinical success rate that is 5% better than another agent. In practice, this number of patients is difficult to obtain in many clinical trials, and their statistical power, therefore, tends to be weak. Data from small clinical trials must always be interpreted carefully. Furthermore, statistically similar outcomes between two antimicrobials in a trial that was designed to show non-inferiority do not mean that the two agents tested have equivalent efficacy-only that the study was not powered to show any difference in efficacy.

The primary problem with clinical trials is that most do not assess antimicrobial efficacy, i.e. the ability of the antimicrobial to eradicate the bacterial infection. Instead, clinical trials generally assess only clinical cure. Data on bacteriologic outcome, if determined, are usually only available from a small number of patients, and the number of antimicrobial-resistant bacterial isolates is often too low to allow the assessment of efficacy. In addition, if included, bacteriologic outcomes may actually be 'presumed bacteriologic success or failure'; that is, the patient had a bacterial pathogen isolated at baseline, and was a clinical cure, or failure, in the absence of a follow-up microbiologically evaluable sample. Another limitation is that, in most trials, patients are enrolled based on a clinical diagnosis, and in the case of respiratory tract infections, patients who do not have a bacterial infection will almost certainly be included. The importance of bacteriologic evaluation in clinical trials cannot be overemphasized.

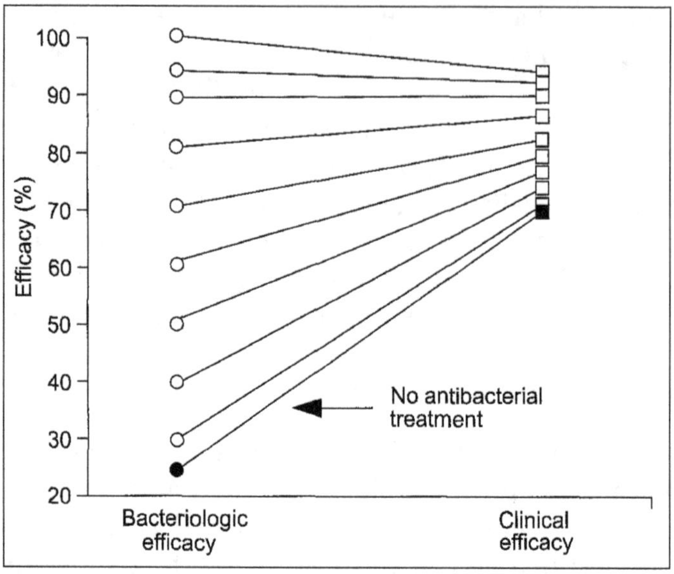

Success of antimicrobial treatment for acute bacterial otitis media based on bacteriologic and clinical criteria.

The perceived value of the clinical assessment of antimicrobials is perpetuated in clinical guidelines. For example, neither the Infectious Diseases Society of America nor the American Thoracic Society guidelines for the management of community-acquired pneumonia indicate that sputum culture is required for outpatients, while both recommend clinical response as a measure of the efficacy of the empirical antimicrobial chosen.

Although clinical assessment may seem to be more practical than bacteriologic evaluation, on its own it may result in a false interpretation of therapeutic success. Marchant et al reviewed the results of randomized, double-blind trials of antimicrobials in acute otitis media to determine bacteriologic and associated clinical success rates. They found that, for a drug with a bacteriologic efficacy of 100%, the calculated clinical efficacy was 93% for acute bacterial otitis media, and for a drug with a bacteriologic efficacy of 27% (equivalent to spontaneous resolution with no antimicrobial therapy), the calculated clinical efficacy was 71%. Thus, an agent that has little antimicrobial activity will appear to be reasonably effective if only clinical criteria are used. The authors of this study named this the 'Pollyanna phenomenon', after the over-optimistic heroine of the eponymous novel.

Primary Goals

The primary goal of antimicrobial therapy should be bacterial eradication. Antimicrobials that maximize bacterial eradication will also maximize clinical cures. The direct association of bacteriologic success and clinical success has been clearly shown by Dagan et al. In their series of 123 patients with acute otitis media, a high clinical success rate (97%) was achieved in those whose middle ear fluid was culture negative on day

4-5 after starting antimicrobial treatment. In contrast, the clinical success rate was only two-thirds of this (63%) in those children whose middle ear fluid was culture positive. A significant association between high bacterial eradication and high clinical cure has also been demonstrated in acute exacerbations of chronic bronchitis.

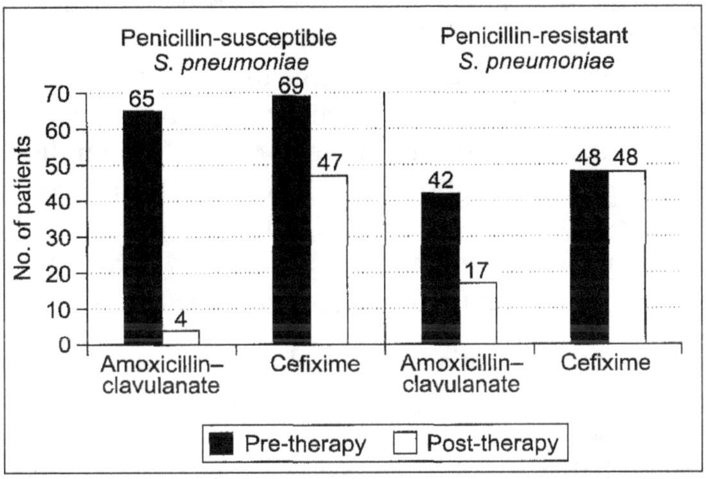

Nasopharyngeal carriage of penicillin-susceptible and penicillin-resistant. Pneumoniae before and after treatment with amoxicillin-clavulanate or cefixime.

Bacteriologic success is also critical to prevent the emergence of resistance. For example, Dabernat et al found that, in nasopharyngeal samples from children with acute otitis media treated with cefixime or amoxicillin-clavulanate, the proportion of S. pneumoniae strains resistant to penicillin was higher at the end of treatment (56.0%, 65/116 S. pneumoniae strains) and at follow-up (50.3%, 69/137) than at inclusion (38.8%) 87/224). Cefixime did not decrease penicillin-resistant S. pneumoniae carriage after treatment, whereas amoxicillin-clavulanate significantly decreased the carriage of these strains, with only 17/42 patients still carrying penicillin-resistant S. pneumoniae after therapy.

The WHO acknowledges the evidence supporting bacterial eradication, issuing the following statement in 2000: 'The most effective strategy against antimicrobial resistance is to get the job done right the first time-to unequivocally destroy microbes-thereby defeating resistance before it starts.' In other words, the old philosophy of saving the best agents for last is wrong, and the most effective antimicrobials should be used from the start to eradicate all bacterial pathogens and thus minimize the selection of resistant strains.

Predicting Bacterial Eradication

If bacterial eradication is the primary goal of antimicrobial therapy, and clinical trials cannot provide us with the data to determine which agents will achieve this goal, how do we proceed towards appropriate prescribing? Studies in both animals and humans indicate that the relationship between pharmacokinetics and pharmacodynamics (PK/PD) can be used to predict antimicrobial bacteriologic efficacy.

Antimicrobials cause either time-dependent killing (e.g. penicillins, cephalosporins, and most macrolides) or concentration-dependent killing (e.g. azalides and fluoroquinolones). For agents causing time-dependent killing, the PK/PD parameter predictive of bacterial eradication is the duration of the dosing interval for which antimicrobial concentrations at the site of infection exceed the MIC of the infecting pathogen, or time above MIC (T > MIC). Mortality in animal models infected with S. pneumoniae is minimized when the T > MIC is 25-40% for penicillins and 40-50% for cephalosporins. For agents, such as the fluoroquinolanes, with concentration-dependent killing, the ratio of the area under the concentration-time curve (AUC) and the MIC, or the ratio of the maximum plasma concentration (C_{max})and the MIC, are predictive of bacterial eradication. For fluoroquinolones, an AUC/MIC ratio of 25 is predictive of bacterial eradication for S. pneumoniae, although ratios of >125 are required to eradicate Gram-negative pathogens, or in immunocompromised patients.

Pharmacodynamic parameters can also be used to predict the potential of an antimicrobial to select resistant strains during therapy. Based on the results of four treatment trials in patients with lower respiratory tract infections treated with fluoroquinolones, Thomas et al calculated that if the AUCYMIC ratio was 100 or higher, only 9.3% of organisms would develop resistance, whereas if the AUC/MIC ratio was less than 100, 82.4% of strains would develop resistance.

The assessment and choice of antimicrobials, the development of new agents and the optimization of currently available agents, should, therefore, be directed by PK/PD predictions of bacterial eradication. By setting a PK/PD target that predicts bacteriologic efficacy against resistant pathogens, doses and regimens can be developed that extend the activity of current agents against these strains. Thus, antimicrobials that can be optimized can continue to provide bacteriologically effective therapy for respiratory tract infections, even in regions where resistance prevalences are high.

EMPIRIC ANTIMICROBIAL THERAPY

Empirical antimicrobial therapy is a reasonable, pragmatic approach to limit the ominous threat of Gram-negative sepsis in febrile neutropenic cancer patients.

Most antibiotic are prescribed without or before knowing the pathogen and its susceptibility to antibiotics. In primary care microbiology diagnostic is usually not available and most patients get well or worse before the microbiology results would come back. At present, extensive microbiology diagnostic in primary care would increase the cost of care with little direct effect on the antibiotic therapy. In hospitals, microbiology diagnostics is encouraged in most patients, but the time to the isolation of a pathogen varies from 12 hours to several days, complete results including susceptibility are almost never available in less than two days. The introduction of rapid molecular diagnostics is urgently needed.

The physician has to decide on antibiotic treatment empirically. Not that far back in the past, in the middle of last century, the physician used to practice “the art of medicine”. There was a general belief that through rigorous medical education, exposure to colleagues and individual experiences each physician always knew and did the right thing. The treatment was litteraly empirical. Later on, in the last forty years, evidence accumulated that the individual physician decisions have to be more informed and based on scientific proofs or at least opinon of colleagues with most experiences in the field. The meaning of the word “empirical”moved from individual experience to “collective” experience and scientific evidence, usually compiled together in various antibiotic treatment guidelines which started to appear in peer reviewed jurnals in the last decade of the 20th century.

In the 21st century empirical antibiotic treatment is challenged again. Antimicrobial resistance ruined many antibiotic treatment regimens proposed in guidelines and used in clinical practice for several decades. Antimicrobial treatment guidelines are becoming more and more tailor-made taking into account antimicrobial resistance situation in the region, in the hospital and even the patient itself. Updates of guidelines are needed to cope with “latest bacterial achievements” in terms of resistance.

Determinants of the Empirical Choice of Antibiotics

Pathogen

Community-acquired infections in immunocompetent patients are caused worldwide by a few micro-organisms. Each anatomic site of infection, i.e. respiratory tract, urinary tract, skin and soft tissue, gastronitestinal tract, genital tract etc. has its typical list of micro-organisms. In respiratory tract infections empirical antibiotic treatment should cover for Streptococcus pneumoniae, in skin and soft tissue empirical treatment should aim to Staphylococcus aureus and Streptococcus pyogenes, in urinary tract infections the terapy should be effective against Escherichia coli. Anatomical diagnosis is therefore crucial for the choice of empiric antibiotic treatment. Age, especially its extremes, may change the spectrum of pathogens in some infections. In bacterial meningitis, Listeria monocytogenes and Enterobacteriaceae play a greater role in neonates and patients above 50 years of age. Broader spectrum of pathogens has to be taken into account in hospital acquired infections and infections in immunocompromised patients. Medical history should include information on travel, contact with resistant micro-organisms, contact to animals, contact with health-care or long term care facilities, previous antibiotic treatment, chronic diseases. Travel to some regions even without health-care contact is a risk factor for colonisation with ESBL positive bacteria and even carbapenem resistance. Health-care contact may be a risk for acquisition of resistant bacteria in some EU countries.

Transmission of methicillin-resistant S. aureus (MRSA) and ESBL producing bacteria in households and between humans and animals has been described several times. Colonisation with resistant bacteria is more common hospitals and in nursing homes than in the community. Virtually hundreds of studies have shown that previous antibiotic treatment is a risk factor for infections/colonisation with resistant micro-organisms. Previous antibiotic treatment may also change the spectrum of micro-organisms. Some chronic diseases predispone to infections with a defined pathogen. Staphyloccal infections are more common in diabetic patients. In patients with chronic obstructive lung disease Haemophilus influenzae pneumonia is more common than in general population and in patients with structural lung disease community-acquired pneumonia may be caused by Pseudomonas aeruginosa, a typical nosocomial pathogen.

Likely Antimicrobial Susceptibility

The susceptibility may be typical for a micro-organism, but may differ from a region to a region, from hospital to hospital and even between hospital wards. Local susceptibility patterns should be provided to prescribers in a timely manner.

Broad-spectrum vs narrow spectrum: several studies have shown the relationship between the appropriate empirical antibiotic tretament and survival in patients with severe sepsis or septic shock, and not in less severely ill patients with bacteremia. Other oucome measures such as length of stay or symptoms durations may be influenced by the appropriateness of the empirical antibiotic choice. The studies in mildly ill patients are lacking. Judicious choice of empirical antibiotic should take into account the possibility of harming the patient with in-appropriate choice and deleterious effect of antimicrobial resistance favourized by broad-spectrum antibiotics for the future patients and infection episodes.

Antimicrobial Dosing

Pharmacokinetical and pharmacodynamic parameters should be taken into account: oral treatment versus parenteral, loading dose is especially important in severely ill patients, the calculation should consider the distribution of antibiotics, the dosing shedule should follow pharmacodynamic parameters.

Antimicrobial Empirical Treatment Check-list

- Anatomical site of infection: potential pathogens.
- Community-acquired vs hospital-acquired/health-care acquired infections.
- Immuno-competent vs immuno-compromised.
- Susceptibility patterns of potential pathogens in the community/hospital.
- Recent hospitalisation.

- Chronic diseases.
- Recent antibiotic treatment.
- Travel.
- Colonisation with resistant bacteria.
- Contact with potential resistant bacteria (human, animal source).
- Severity of the infection/vulnerable patient.
- The first dose of antibiotic: volume of antibiotic distribution, body mass.
- The dosing interval, maintenance dosing: pharmacodynamic of the antibiotic, excretion problems (kidney function etc).

Does Appropriate Empirical Treatment Matter?

There is a general belief that bacteria should be killed or at least stop multiplying to cure the infection. Therefore the choice of antibiotics which are not effective against the pathogen should have serious consequences for the patient. Indeed several studies have proven that appropriate treatment is related to improvement and/or survival.

First studies investigating the relevance of appropriate antibiotic treatment were published twenty years ago and focused on most severely ill patients in intensive care units (ICU).

A large systematic review of studies looking at the relationship between appropriate antimicrobial therapy and mortality in patients with sepsis including 48 studies providing data on adjusted mortality showed that in-appropriate is related to higher mortality. In spite of heterogenicity of the studies and variable results, the relationship between appropriate antibiotic therapy and survival was visible in subgroups of patients regarding age, neutropenia, pathogen groups, presence of bacteremia, and source of infection. The impact of appropriate antibiotic treatment was greater in patients presenting with septic shock than in less severely ill patients.

Empirical Treatment: How Fast?

The choice of empirical antibiotic is a complex decision which should be done as soon as possible. The definition of the time-frame for antibiotic prescribing depends upon the type of infection and is not well defined in many situations.

In acute bacterial meningitis, the studies investigating the role of pre-hospital administration of antibiotics did not show a convincing benefit. The role of pre-hospital antibiotics depends upon the individual patient course of the disease. In some patients it may take a few days untilf the typical clinical picture of acute bacterial meninigitis develops, in the others, the course is much more fulminant. The door-to needle time is better defined, it should not exceed 3 to 6 hours.

In community-acquired pneumonia two restrospective studies showed that the delay of antibiotic administration eight and four hours beyond the admission is related to higher mortality. More recent prospective studies did not confirm their results. Faster administration of antibiotics did not correlate with the stabilisation of the patient, but may be related to shorter length of hospital stay. In recent guideliness the recommendation for administration of antibiotics within a defined time-frame after admission was ommited, and replaced by the administration of antibiotics just after the diagnosis of pneumonia thus avoiding misdiagnosis and unnecessary antibiotic treatment.

In a large retrospective study in patients with septic shock the survival was significally higher in patients who received antibiotics in the first hour of documented hypotension. In the international guidelines for the management of severe sepsis and septic shock (Surviving Sepsis Campaign), the recommendation of antibiotic administration within the first hour was extended from septic shock to severe sepsis.

There are less data on the timing of antibiotic treatment in less severe infections. In children, the Dutch approach named "wait-and-watch" was described in acute otitis media in children 2 to 12 years of age. The approach does not favor timely antibiotic treatment but a delay of three to four days in which most children get better with symptomatic treatment thus avoiding excessive use of antibiotics. More recent meta-analysis showed that observational strategy is not beneficial in children younger than 2 years and bilateral otitis and children with acute otitis and otorrhoea.

Table: "Door-to-needle" times in severe infections.

Syndrome	Interval after admission in which empirical antibiotic treatment should be administered
Acute bacterial meningitis	< 3 to 6 hours
Community-acquired pneumonia	Immediatelly after the diagnosis
Septic shock, severe sepsis	< 1 hour

ANTIMICROBIAL CHEMOTHERAPY

Various antimicrobial agents act by interfering with (1) cell wall synthesis, (2) plasma membrane integrity, (3) nucleic acid synthesis, (4) ribosomal function, and (5) folate synthesis.

Action of Specific Agents

Cell wall synthesis is inhibited by ß-lactams, such as penicillins and cephalosporins, which inhibit peptidoglycan polymerization, and by vancomycin, which combines with cell wall substrates. Polymyxins disrupt the plasma membrane, causing leakage. The plasma membrane sterols of fungi are attacked by polyenes (amphotericin) and imidazoles. Quinolones bind to a bacterial complex of DNA and DNA gyrase, blocking DNA replication. Nitroimidazoles damage DNA. Rifampin blocks RNA synthesis by binding to DNA directed RNA polymerase. Aminoglycosides, tetracycline, chloramphenicol, erythromycin, and clindamycin all interfere with ribosome function. Sulfonamides and trimethoprim block the synthesis of the folate needed for DNA replication.

Bacterial Resistance

Bacteria can evolve resistance to antibiotics. Resistance factors can be encoded on plasmids or on the chromosome. Resistance may involve decreased entry of the drug, changes in the receptor (target) of the drug, or metabolic inactivation of the drug.

Effects of Combination Therapy

Combinations of antibiotics may act synergistically-producing an effect stronger than the sum of the effects of the two drugs alone or antagonistically, if one agent inhibits the effect of the other.

Adverse Effects of Antimicrobial Agents

Many antibiotics are toxic to the host. Alterations of the normal intestinal flora caused by antibiotics may result in diarrhea or in superinfection with opportunistic pathogens.

Biochemical Basis of Antimicrobial Action

Bacterial cells grow and divide, replicating repeatedly to reach the large numbers present during an infection or on the surfaces of the body. To grow and divide, organisms must synthesize or take up many types of biomolecules. Antimicrobial agents interfere with specific processes that are essential for growth and/or division. They can be separated into groups such as inhibitors of bacterial and fungal cell walls, inhibitors of cytoplasmic membranes, inhibitors of nucleic acid synthesis, and inhibitors of ribosome function. Antimicrobial agents may be either bactericidal, killing the target bacterium or fungus, or bacteriostatic, inhibiting its growth. Bactericidal agents are more effective, but bacteriostatic agents can be extremely beneficial since they permit the normal defenses of the host to destroy the microorganisms.

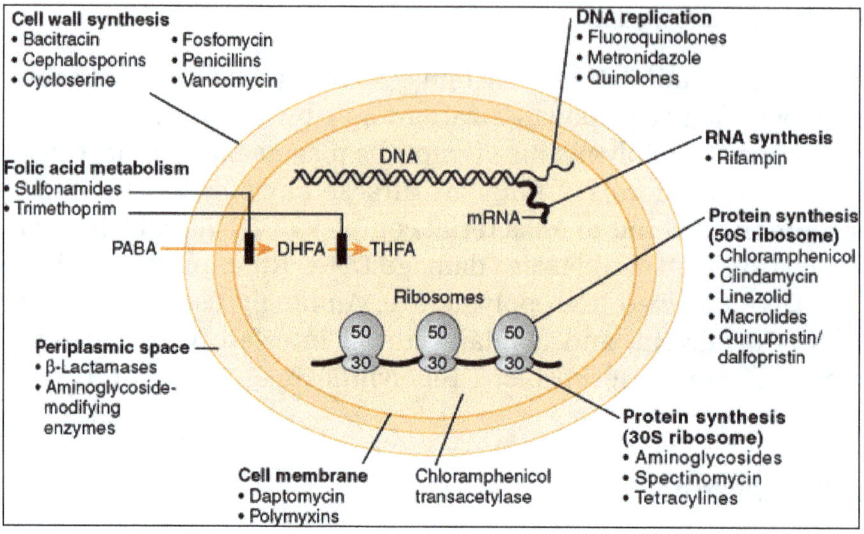

Sites of action of different antimicrobial agents. PABA, paraminobenzoic acid; DHFA, dihydrofolic acid; THFA, tetrahydrofolic acid.

Inhibition of Bacterial Cell Wall Synthesis

bacteria are classified as Gram-positive and Gram-negative organisms on the basis of staining characteristics. Gram-positive bacterial cell walls contain peptidoglycan and teichoic or teichuronic acid, and the bacterium may or may not be surrounded by a protein or polysaccharide envelope. Gram-negative bacterial cell walls contain peptidoglycan, lipopolysaccharide, lipoprotein, phospholipid, and protein. The critical attack site of anti-cell-wall agents is the peptidoglycan layer. This layer is essential for the survival of bacteria in hypotonic environments; loss or damage of this layer destroys the rigidity of the bacterial cell wall, resulting in death.

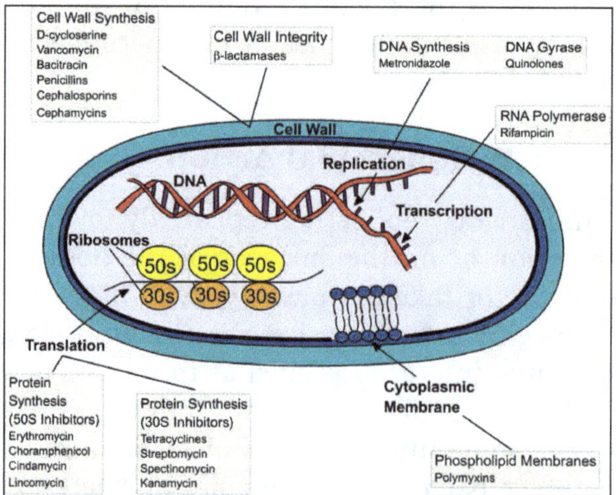

Outer wall of Gram-positive and Gram-negative species and detail of porin channels of Gram-negative bacteria. Antimicrobial agents diffuse easily through the loose outer wall of Gram-positive bacteria, but must go through the narrow channels of the Gram-negative species.

Peptidoglycan synthesis occurs in three stages. The first stage takes place in the cytoplasm, where the low-molecular-weight precursors UDP-GlcAc and UDP-MurNAc-L-Ala-D-Glu-meso-Dap-D-Ala-D-Ala are synthesized. A number of antimicrobial agents interfere with these early steps in cell wall biosynthesis. UTP and N-acetylglucosamine α-1-P are converted to UDP-N-acetylglucosamine, which is subsequently converted by the enzyme phosphoenolpyruvate: UDP-GIcNAc-3-enol-pyruvyltransferase. Fosfomycins block this transfer by a direct nucleophilic attack on the enzyme. Because mammalian enzymes such as enolase, pyruvate kinase, carboxykinases, and the shikimate enlaces are not inhibited by these compounds, fosfomycins have no effect on the host. Three amino acids are added to the muramyl peptide to yield a tripeptide to which two more amino acids will be linked. The dipeptide D-alanyl-D-alanine is synthesized from two molecules of D-alanine by the enzyme D-alanyl-D-alanine synthetase. D-Alanine is produced from L-alanine by an alanine racemase. Cycloserine inhibits both alanine racemase and D-alanyl-D-alanine synthetase owing to the structural similarity of cycloserine and D-alanine and to the fact that cycloserine actually binds to the enzymes better than the D-alanine.

The second stage of cell wall synthesis is catalyzed by membrane-bound enzymes. The nonnucleotide portion of the precursor molecules previously made are transferred sequentially to a carrier in the cytoplasmic membrane. This carrier is a phosphorylated undecaprenyl alcohol. The lipid carrier functions as a point of attachment to the membrane for the precursors and allows for transport of the subunits across the hydrophobic interior of the cytoplasmic membrane to the outside surface. Bacitracin is a peptide antibiotic that specifically interacts with the pyrophosphate derivate of the undecaprenyl alcohol, preventing further transfer of the muramylpentapeptide from the precursor nucleotide to the nascent peptidoglycan.

The third stage of cell wall synthesis involves polymerization of the subunits and the attachment of nascent peptidoglycan to the cell wall. Polymerization occurs by transfer of the new peptidoglycan chain from its carrier in the membrane to the nonreducing N-acetylglucosamine of the new saccharide-peptide that is attached to the membrane. The new peptidoglycan is attached to preexisting cell wall peptidoglycan by a transpeptidase reaction that involves peptide chains in both polymers, one of which must possess a D-alanyl-D-alanine terminus. It is believed that the transpeptidase enzyme cleaves the peptide bond between two D-alanyl residues in the pentapeptide and become acylated via the carbonyl group of the penultimate D-alanine residue. This final reaction is inhibited by β-lactam antibiotics. These antibiotics contain a critical four-membered ring, which undergoes an acylation reaction with the transpeptidases that cross-link the polymers. The β-lactam antibiotics are the penicillins (penams), cephalosporins (including oxacephems and cephamycins), penems, thienamycins (carbapenems), and aztreonam (monobactams). The enzymes involved in this final process of cell wall formation are called penicillin-binding proteins since they were discovered by labeling with radioactive penicillin G. The enzymes are different in Gram-positive and Gram-negative bacteria and in anaerobic species. Differences in the penicillin-binding proteins explain, to some extent, differences in antibacterial activity of the β-lactam antibiotics. The penicillin-binding protein, to which a particular β-lactam antibiotic binds, affects the morphologic response of the bacterium to the agent. For example, some antibiotics bind to a penicillin-binding protein that is involved in forming the septum between dividing cells; as a result, the bacteria continue to grow into long filaments, which eventually die. Binding to another penicillin-binding protein results in rapid lysis of a bacterium because the wall bulges and the bacterium bursts. β-Lactams such as mecillinam (an amidino penicillin) do not bind to the penicillin-binding proteins of Gram-positive bacteria and therefore do not affect these bacteria. Aztreonam binds only to Gram-negative penicillin-binding proteins and does not inhibit Gram-positive or anaerobic species.

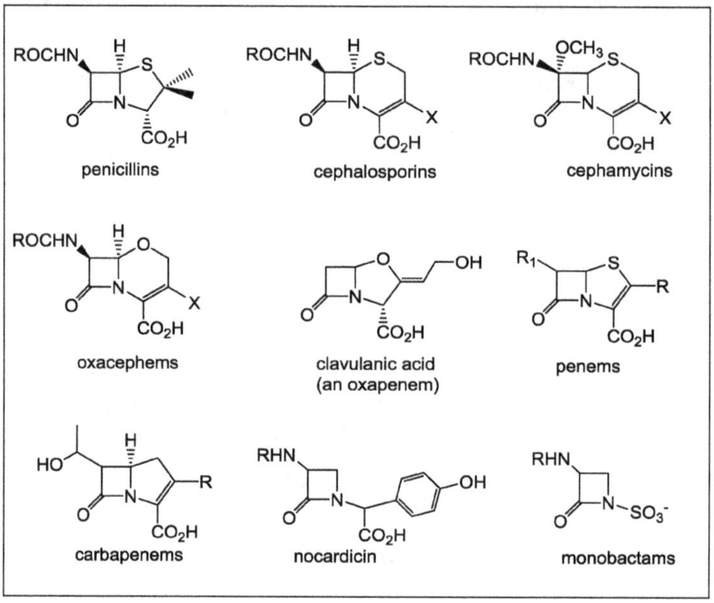

Basic structures of β-lactam antibiotics. Penicillins and cephalosporins/cephamycins are widely used to inhibit both Gram-positive and Gram-negative bacilli. Monobactams inhibit only aerobic Gram-negative bacilli, clavulanic acid acts as a β-lactamase inhibitor, and thienamycin inhibits a wide range of aerobic and anaerobic species. R and R[1] represent various carbon groups. X can be either hydrogen or a mehoxy group.

Vancomycin interrupts cell wall synthesis by forming a complex with the C-terminal D-alanine residues of peptidoglycan precursors. Complex formation at the outer surface of the cytoplasmic membrane prevents the transfer of the precursors from a lipid carrier to the growing peptidoglycan wall by transglycosidases. Biochemical reactions in the cell wall catalyzed by transpeptidases and D,D-carboxypeptidases are also inhibited by vancomycin and other glycopeptide antimicrobials. Because of its large size and complex structure, vancomycin does not penetrate the outer membrane of gram-negative organisms. With resistance to beta-lactams increasing in frequency among staphylococci and enterococci, glycopeptides such as vancomycin remain important therapeutic agents against such bacteria.

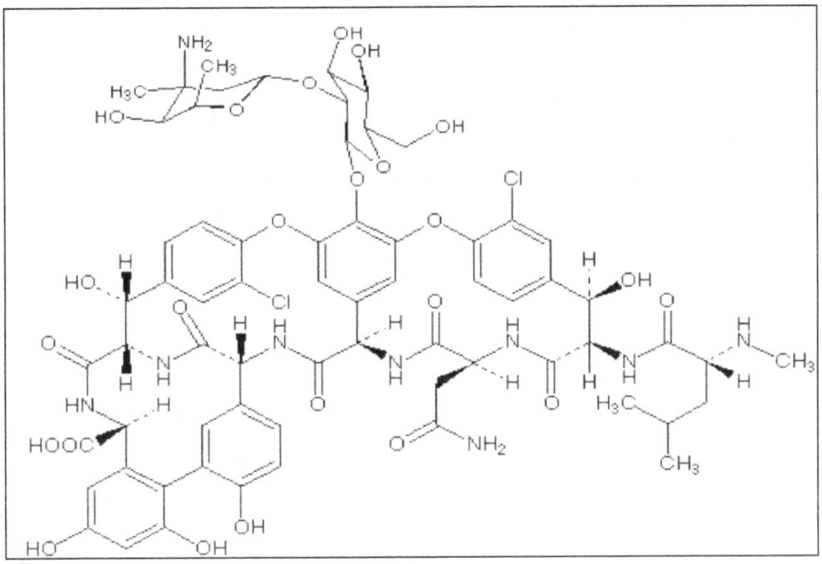

Structure of vancomycin.

Antibiotics that Affect the Function of Cytoplasmic Membranes

Bacterial Cytoplasmic Membranes

Biologic membranes are composed basically of lipid, protein, and lipoprotein. The cytoplasmic membrane acts as a diffusion barrier for water, ions, nutrients, and transport systems. Most workers now believe that membranes are a lipid matrix with globular proteins randomly distributed to penetrate through the lipid bilayer. A number of antimicrobial agents can cause disorganization of the membrane. These agents can be divided into cationic, anionic, and neutral agents. The best-known compounds are polymyxin B and colistemethate (polymyxin E). These high-molecular-weight octapeptides inhibit Gram-negative bacteria that have negatively charged lipids at the surface. Since the activity of the polymyxins is antagonized by Mg^{2+} and Ca^{2+}, they probably competitively displace Mg^{2+} or Ca^{2+} from the negatively charged phosphate groups on membrane lipids. Basically, polymyxins disorganize membrane permeability so that nucleic acids and cations leak out and the cell dies. The polymyxins are of virtually no use as systemic agents since they bind to various ligands in body tissues and are potent toxins for the kidney and nervous system. Gramicidins are also membrane-active antibiotics that appear to act by producing aqueous pores in the membranes. They also are used only topically.

Fungal Membranes

Fungal membranes contain sterols, whereas bacterial membranes do not. The polyene antibiotics, which apparently act by binding to membrane sterols, contain a rigid hydrophobic center and a flexible hydrophilic section. Structurally, polyenes are tightly packed rods held in rigid extension by the polyene portion. They interact with fungal cells to produce a membrane-polyene complex that alters the membrane permeability, resulting in internal acidification of the fungus with exchange of K^+ and sugars; loss of phosphate esters, organic acids, nucleotides; and eventual leakage of cell protein. In effect, the polyene makes a pore in the fungal membrane and the contents of the fungus leak out. Prokaryotic cells neither bind to nor are inhibited by polyenes. Although numerous polyene antibiotics have been isolated, only amphotericin B is used systemically. Nystatin is used as a topical agent and primaricin as an ophthalmic preparation.

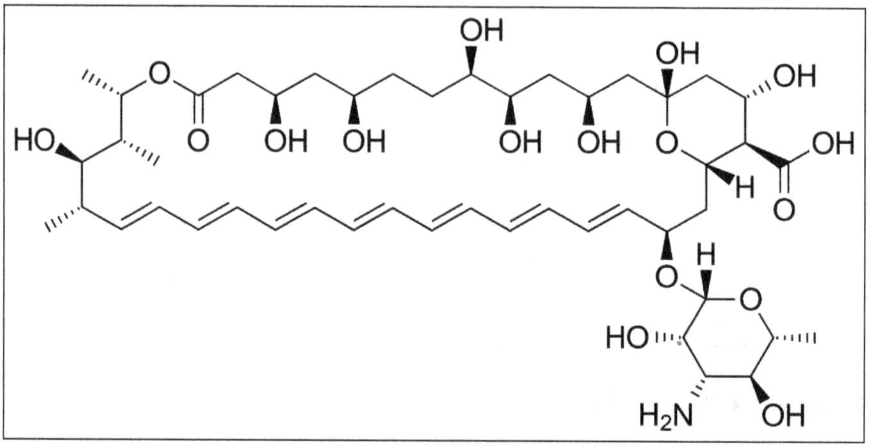

Structure of amphotericin B.

A number of other agents interfere with the synthesis of fungal lipid membranes. These agents belong to a class of compounds referred to as imidazoles: miconazole, ketoconazole, clotrimazole, and fluconazole. These compounds inhibit the incorporation of subunits into ergosterol and may also directly damage the membrane.

Antibiotics that Inhibit Nucleic Acid Synthesis

Antimicrobial agents can interfere with nucleic acid synthesis at several different levels. They can inhibit nucleotide synthesis or interconversion; they can prevent DNA from functioning as a proper template; and they can interfere with the polymerases involved in the replication and transcription of DNA.

Interference with Nucleotide Synthesis

A large number of agents interfere with purine and pyrimidine synthesis or with the interconversion or utilization of nucleotides. Other agents act as nucleotide analogs that are incorporated into polynucleotides.

Flucytosine (5-fluorocytosine) is an antifungal agent that inhibits yeast species. It is converted in the fungal cell to 5-fluorouracfl, which inhibits thymidylate synthetase resulting in a deficit of thymine nucleotides and impaired DNA synthesis. Adenosine arabinoside inhibits viruses. It is phosphorylated in virus-infected cells and acts as a competitive analog of DATP, inhibiting the incorporation of DATP into DNA. Acyclovir is a nucleoside analog that, after being converted to a triphosphate, inhibits the thymidine kinase and DNA polymerase of herpes viruses. Zidovudine (AZT) inhibits human immunodeficiency virus (HM replication by interfering with viral RNA-dependent DNA polymerase (reverse transcriptase).

Agents That Impair the Template Function of DNA

A number of substances bind to DNA by intercalation. None of them is useful as an antibacterial agent; however, chloroquine and miracil D (lucanthone) inhibit plasmodia and schistosomes, respectively. These agents are thought to intercalate into the DNA and thereby to inhibit further nucleic acid synthesis. Acridine dyes such as proflavine act by this intercalation mechanism, but because they are toxic and carcinogenic in mammals they cannot be used as antibacterial agents.

Inhibition of DNA-directed DNA Polymerase

Rifamycins are a class of antibiotics that inhibit DNA-directed RNA polymerase. Polypeptide chains in RNA polymerase attach to a factor that confers specificity for the recognition of promoter sites that initiate transcription of the DNA. Rifampin binds noncovalently but strongly to a subunit of RNA polymerase and interferes specifically with the initiation process. However, it has no effect once polymerization has begun.

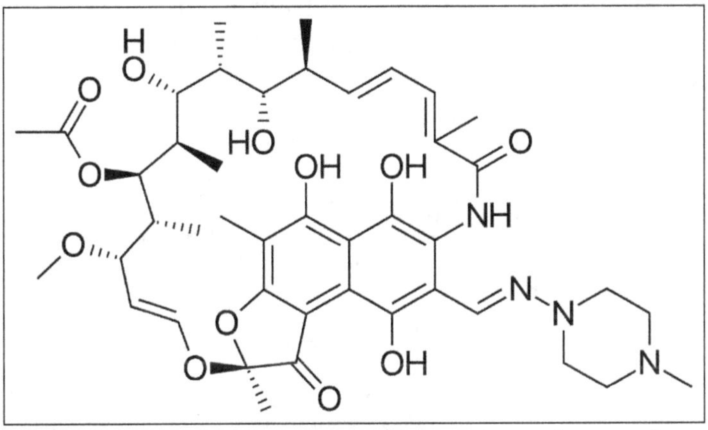

6 Structure of rifampin, which inhibits the DNA-directed RNA polymerase.

Inhibition of DNA Replication

DNA gyrase and topoisomerase I act in concert to maintain an optimum supercoiling state of DNA in the cell. In this capacity, DNA gyrase is essential for relieving torsional strain during replication of circular chromosomes in bacteria. The enzyme is a tetrameric protein composed of two A and two B subunits. A transient, covalent bond between the A subunit and DNA occurs during the double strand passage reaction catalyzed by gyrase. Quinolones such as nalidixic acid, bind to the cleavage complex composed of

DNA and gyrase during this strand passage. This interaction of quinolone acts to stabilize the cleavage intermediate which has a detrimental effect on the normal DNA replication process. The effects of this inhibition result in the death of the bacterial cell. The newer fluoroquinolones such as ciprofloxacin, norfloxacin, and ofloxacin also interact with DNA gyrase and possess a broad spectrum of antimicrobial activity.

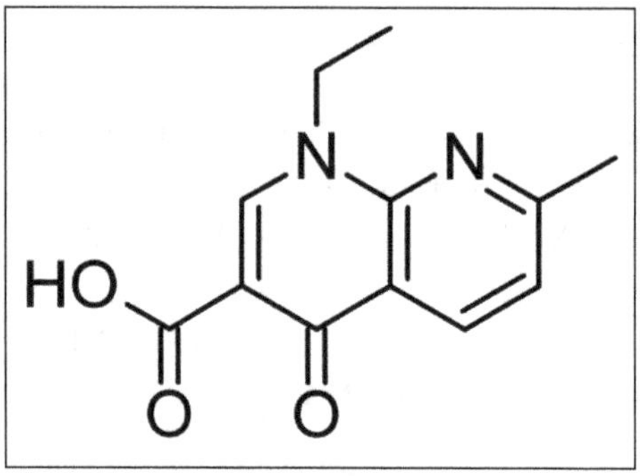

Nalidixic Acid.

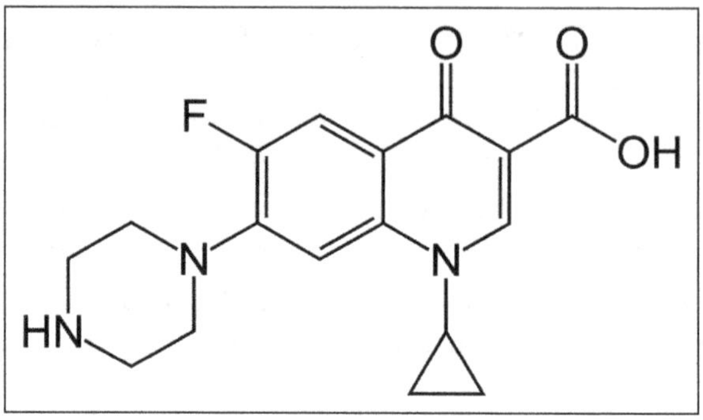

Ciprofloxacin Acid.

Structure of quinolone antibiotic. Nalidixic inhibits only aerobic Gram-negative species. In ciprofloxicin, the flourine provides Gram-positive activity, the piperazine group increases activity against members of the Enterovacteriaceae, and the piperazine and cylopropyl groups give activity against Pseudomonas species.

Nitroimidazoles such as metronidazole inhibit anaerobic bacteria and protozoa. The nitro group of the nitrosohydroxyl amino moiety is reduced by an electron transport protein in anaerobic bacteria. The reduced drug causes strand breaks in the DNA. Mammalian cells are unharmed because they lack enzymes to reduce the nitro group of these agents.

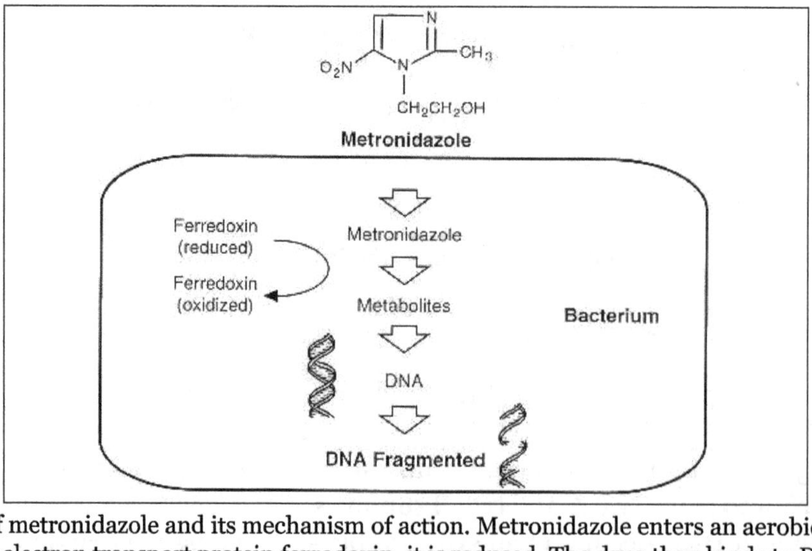

Structure of metronidazole and its mechanism of action. Metronidazole enters an aerobic bacterium where, via the electron transport protein ferrodoxin, it is reduced. The drug then binds to DNA, and DNA breakage occurs.

Antimicrobial Inhibitors of Ribosome Function

The basic structure and function of ribosomes are presented in figure. A number of antibacterial agents act by inhibiting ribosome function. Bacterial ribosomes contain two subunits, the 50S and 30S subunits, and it is possible to localize the action of antibiotics to one or both subunits. It is also possible to isolate the specific ribosomal proteins to which an agent binds and to isolate bacterial mutants that lack a specific ribosomal protein and therefore show resistance to a particular agent.

Aminoglycosides act by binding to specific ribosomal subunits. Aminoglycosides are complex sugars connected in glycosidic linkage. They differ both in the molecular nucleus, which can be streptidine or 2-deoxystreptidine, and in the aminohexoses linked to the nucleus. Essential to the activity of these agents are free NH4 and OH groups by which aminoglycosides bind to specific ribosomal proteins. Streptomycin, the first aminoglycoside studied, was a useful too] in elucidating protein synthesis. However, it is rarely used clinically today except to treat tuberculosis, and its mode of action differs to some extent from that of the other clinically useful aminoglycosides, which are 2-deoxystreptidine derivatives such as gentamicin, tobramycin, and amikacin. Streptomycin binds to a specific S12 protein in the 30S ribosomal subunit and causes the ribosome to misread the genetic code. Other aminoglycosides bind not only to the S12 protein of the 30S ribosome, but also to some extent to the L6 protein of the 50S ribosome. This latter binding is quite important in terms of the resistance of bacteria to aminoglycosides. Indeed, the aminoglycoside-type drugs can combine with other binding sites on 30S ribosomes, and they kill bacteria by inducing the formation of aberrant, nonfunctional complexes as well as by causing misreading.

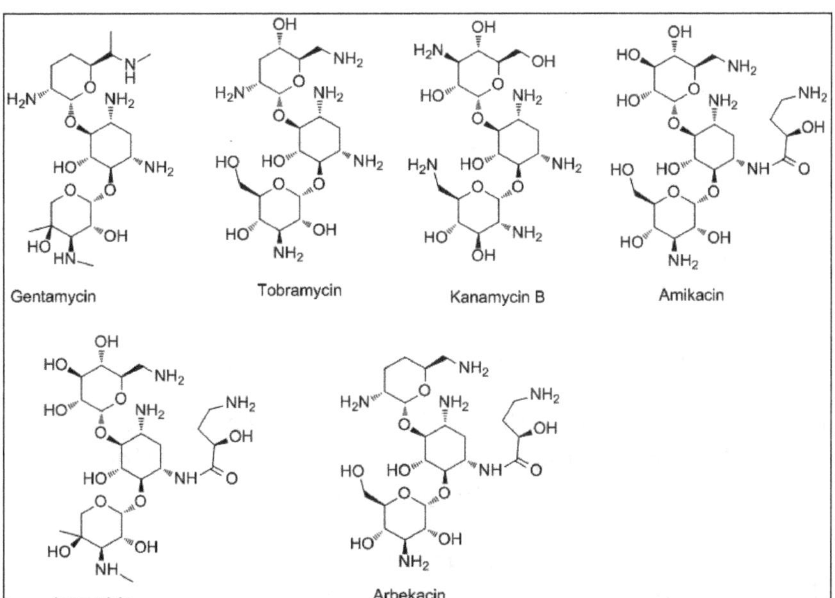

Structures of three aminoglycoside antibiotics used clinically. Critical aspectos of the molecules are the amino and hydroxy groups that bind to proteins in the ribosomes.

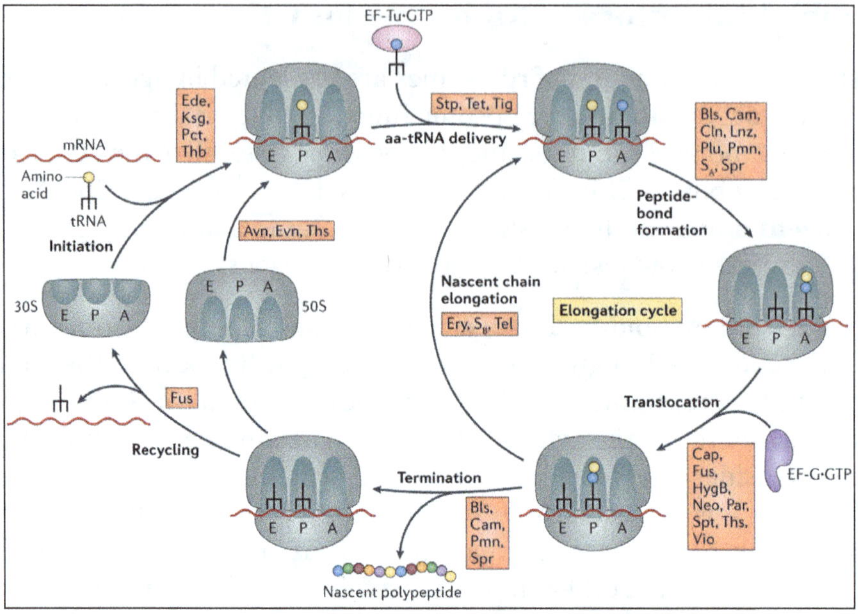

Diagrammatic representation of inhibition sites of protein biosynthesis by various antibiotics that bind to the 30S and 50S ribosomes.

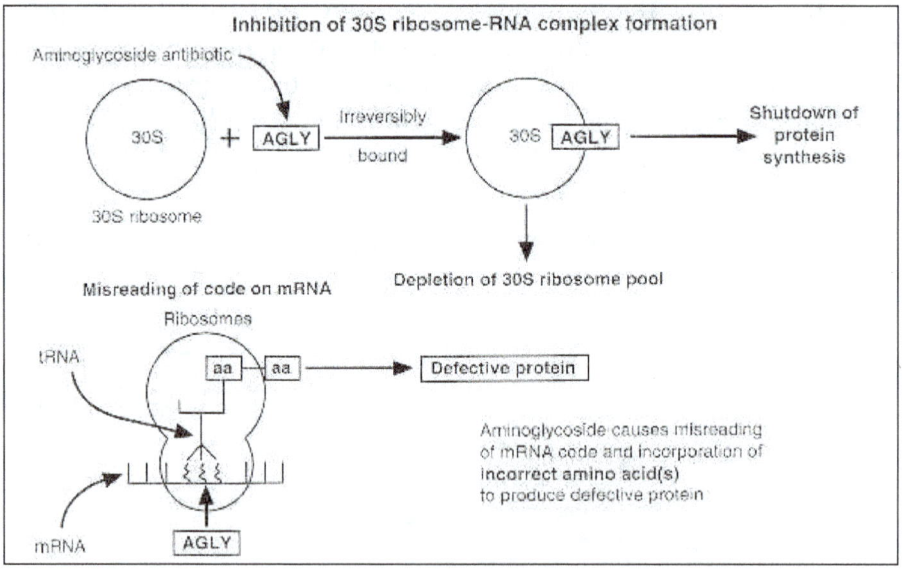

Inhibition of protein biosynthesis by aminoglycosides.

Spectinomycin is an aminocylitol antibiotic that is closely related to the aminoglycosides. It binds to a different protein in the ribosome and is bacteriostatic but not bactericidal. It is used to treat penicillin-resistant gonorrhea.

Other agents that bind to 30S ribosomes are the tetracyclines. These agents appear to inhibit the binding of aminoacyl-tRNA into the A site of the bacteria] ribosome. Tetracycline binding is transient, so these agents are bacteriostatic. Nonetheless, they inhibit a wide variety of bacteria, chlamydias, and mycoplasmas and are extremely useful antibiotics.

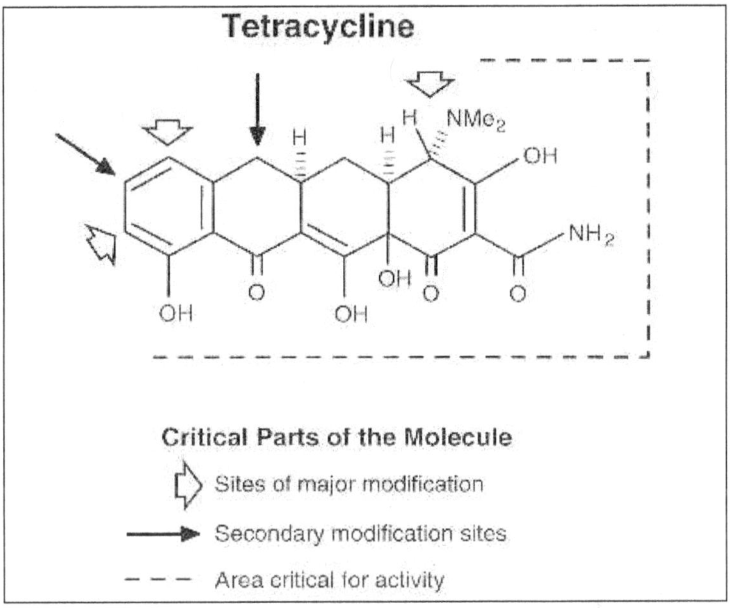

Structure of tetracycline showing the area critical for activity and major and minor points of modification.

There are three important classes of drugs that inhibit the 50S ribosomal subunit. Chloramphenicol is a bacteriostatic agent that inhibits both Gram-positive and Gram-negative bacteria. It inhibits peptide bond formation by binding to a peptidyltransferase enzyme on the 50S ribosome. Macrolides are large lactone ring compounds that bind to 50S ribosomes and appear to impair a peptidyltransferase reaction or translocation, or both. The most important macrolide is erythromycin, which inhibits Gram-positive species and a few Gram-negative species such as Haemophilus, Mycoplasma, Chlamydia, and Legionella.

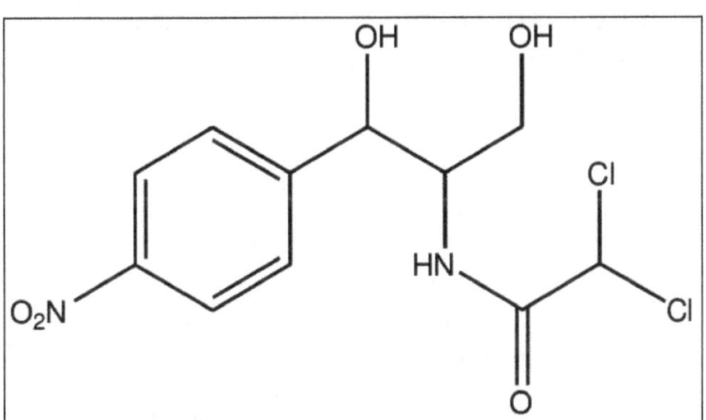

Structure of chloramphenicol.

New molecules such as azithromycin and clarithromycin have greater activity than erythromycin against many of these pathogens. Lincinoids, of which the most important is clindamycin, have a similar site of activity. Both macrolides and lincinoids are generally bacteriostatic. inhibiting only the formation of new peptide chains.

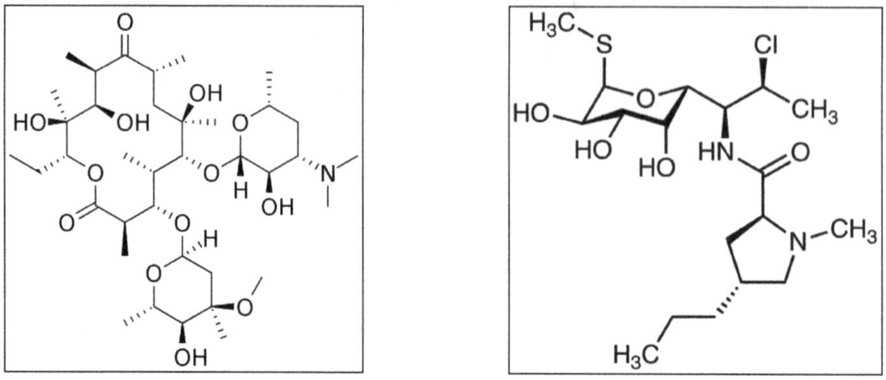

Structure of erythromycin (prototype or macrolide) and clindamycin. Although extremely different in structure, both compounds inhibit protein synthesis by binding to 50S ribosome.

Drugs that Inhibit other Biochemical Targets

Both trimethoprim and the sulfonamides interfere with folate metabolism in the bacterial cell by competitively blocking the biosynthesis of tetrahydrofolate, which acts as a carrier of one-carbon fragments and is necessary for the ultimate synthesis of DNA, RNA and bacterial cell wall proteins. Unlike mammals, bacteria and protozoan parasites usually lack a transport system to take up preformed folic acid from their environment. Most of these organisms must synthesize folates, although some are capable of using exogenous thymidine, circumventing the need for folate metabolism.

Sulfonamides competitively block the conversion of pteridine and p-aminobenzoic acid (PABA) to dihydrofolic acid by the enzyme pteridine synthetase. Sulfonamides have a greater affinity than p-aminobenzoic acid for pteridine synthetase. Trimethoprim has a tremendous affinity for bacterial dihydrofolate reductase (10,000 to 100,000 times higher than for the mammalian enzyme); when bound to this enzyme, it inhibits the synthesis of tetrahydrofolate.

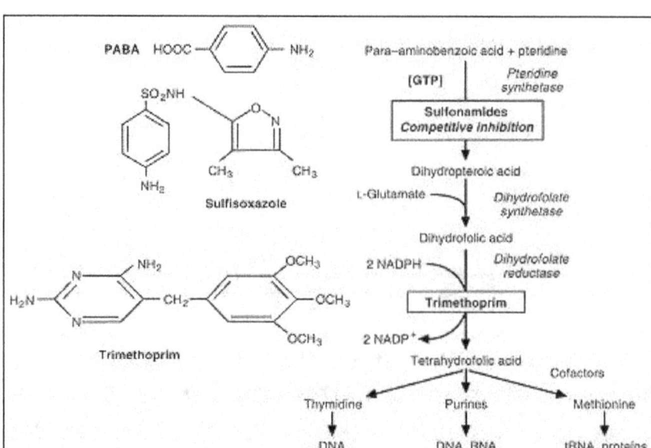

Structure of sulfonamide and trimethoprim with sites of inhibition of folic metabolism.

Antibacterial Agents that Affect Mycobacteria

Isoniazid is a nicotinamide derivative that inhibits mycobacteria. Its precise mode of action is not known, but it affects the synthesis of lipids, nucleic acids, and the mycolic acid of the cell walls of these species. Ethambutol is also an antimycobacterial agent whose mechanism of action is unknown. It is mycostatic, whereas isoniazid is mycocidal. The other antituberculosis drugs, rifampin and streptomycin, affect mycobacteria in the same manner that they inhibit bacteria. Pyrazinamide is a synthetic analog of nicotinamide. It is bactericidal, but its exact mechanism is unknown.

Bacterial Resistance

Bacteria have proved adept at developing resistance to new antimicrobial agents. There are a number of ways in which bacteria can become resistant. Most of the early studies of bacterial resistance focused on single-step mutational events of chromosomal origin. Resistance to the early sulfonamides, for example, was the result of a single amino acid change in the enzyme pteridine synthetase that caused sulfonamides to bind less well than p-aminobenzoic acid. Similarly, a single step mutation that altered a ribosomal protein conferred resistance to streptomycin. In the late 1950s, Japanese workers found that enteric bacteria such as Shigella dysenteriae had become resistant not only to sulfonamides but also to the tetracyclines and chloramphenicol. This resistance was due not to a chromosomal change, but rather to the presence of extrachromosomal DNA that was transmissible. This type of resistance is called plasmid-mediated resistance.

Table: Mechanisms of resistance.

Alteration of target:	Modification to insensitivity to inhibitor. Reduction in physiologic importance of target. Synthesis of new modified drug to enter cell.
Prevention of access to target:	Efflux of more drug than enters cell. Failure of modified drug to enter cell.
Inactivation of agent:	Destruction of the agent. Modification of the agent so it fails to bin to target.
	Failure to convert an inactive precursor agent to its active form.

Resistance-conferring plasmids are present in virtually all bacteria. For example, resistance to ampicillin appeared in Haemophilus influenzae in 1974 and in Neisseria gonorrhoeae in 1976. In the last several years, organisms such as enterococci have been shown to contain plasmids that confer resistance to drugs such as ampicillin and aminoglycosides.

Bacteria also contain transposons, which can insert into plasmids and also into the chromosome. Transposon-mediated resistance to most of the major antibiotics has been found in the past few years.

Antimicrobial agents exert a strong selective pressure on the development of both chromosomal and plasmid-mediated resistance. Administration of an antibiotic destroys the susceptible bacteria in a population, but may permit resistant ones to proliferate. From an epidemiologic viewpoint, plasmid-mediated resistance is the most important type, since it is transmissible, is usually highly stable, confers resistance to many different classes of antibiotics simultaneously, and often is associated with other characteristics that enable a microorganism to colonize and invade a susceptible host.

Mechanisms of Resistance

The basic mechanisms by which a microorganism can resist an antimicrobial agent are (1) to alter the receptor for the drug (the molecule on which it exerts its effect); (2) to decrease the amount of drug that reaches the receptor by altering entry or increasing removal of the drug; (3) to destroy or inactivate the drug; and (4) to develop resistant metabolic pathways. Bacteria can possess one or all of these mechanisms simultaneously.

Resistance due to Altered Receptors

β-lactam Resistance

The ability to analyze changes in receptors for β-lactams by competition experiments in which [14C] penicillin is inhibited from binding to penicillin-binding proteins has explained a number of cases of bacterial resistance to penicillins and cephalosporins. In 1977, Streptococcus pneumoniae strains resistant to penicillin G were encountered in South Africa. Plasmids were not the cause of the resistance. Penicillin-resistant S pneumoniae cells have altered penicillin-binding proteins, which bind penicillin less well. Resistance of S pneumoniae to penicillin has been increasing, and there are now relatively resistant isolates (minimal inhibitory concentration [MIC], 0.1 to 1 mg/ml) in many parts of the world.

Altered penicillin-binding proteins also explain the resistance of some Staphylococcus aureus strains to β-lactamase-stable penicillins (the so-called methicillin-resistant strains). The β-lactams induce synthesis of a new penicillin-binding protein, PBP2a, which does not bind any β-lactam. The β-lactam resistance of coagulase-negative staphylococci is also the result of altered penicillin-binding proteins. Staphylococcal organisms resistant to methicillin are resistant to all penicillins, cephalosporins, and carbapenems.

The resistance of group D streptococci to β-lactam antibiotics appears to be the result of lower affinity of the penicillin-binding proteins for the penicillins. Enterococci are resistant to all cephalosporins because of failure to bind to the penicillin-binding proteins. One Gram-negative species for which resistance to β-lactam antibiotics can be correlated with diminished affinity of the target enzymes is N gonorrhoeae.

Vancomycin Resistance

Certain transposable genetic elements encode special cell wall-synthesizing enzymes which change the structure of the normal D-Ala-D-Ala side chain in the peptidoglycan assembly pathway. The altered side chain (D-Ala-D-Lac) does not bind vancomycin and allows normal peptidoglycan polymerization to occur in the presence of the drug. Depending upon the nature of the vancomycin resistance gene, high-level resistance can occur to glycopeptides. Thus far, this type of resistance has been found in enterococci but not in multi-resistant isolates of Staphylococcus aureus.

Macrolide-lincomycin Resistance

Macrolide-lincomycin resistance in clinical isolates of staphylococci and streptococci has been recognized for several decades. The resistance is due to methylation of two adenine nucleotides in the 23S component of 50S RNA. This resistance is plasmid mediated, and the resistance is encoded on transposons. Resistance results from induction of an enzyme that is normally repressed. The methylated RNA binds macrolidelincomycin-type drugs less well than unmethylated RNA does. Induction of resistance varies by species, and in most Gram-positive species erythromycin is a more effective inducer of resistance than is clindamycin. The plasmids that mediate macrolide-lincomycin resistance in streptococci and staphylococci have extensive structural similarity, indicating that these plasmids readily pass between these species.

Rifampin Resistance

The resistance of bacteria to rifampin is caused by an alternation of one amino acid in DNA-directed RNA polymerase, which results in reduced binding of rifampin. The degree of resistance is related to the degree to which the enzyme is changed, but does not correlate strictly with enzyme inhibition. This form of resistance occurs at a low level in any population of bacteria so that resistance develops by natural selection during a course of therapy. Naturally resistant organisms are more common among members of the Enterobacteriaceae, explaining why agents of urinary tract infections rapidly became resistant to rifampin. The resistance of Neisseria meningitides to rifampin appeared in closed military settings in which rifampin has been used for prophylaxis.

Sulfonamide-trimethoprim Resistance

Sulfonamide can be rendered ineffective by altered or new dihydropteroic synthetase that has poor affinity for sulfonamides and preferentially binds p-aminobenzoic acid. Sulfonamide resistance of this type can result from a point mutation or from acquisition of a plasmid that causes synthesis of the new enzyme. A most serious resistance problem is an increase in resistance to trimethoprim. This plasmid- and transposon-mediated resistance is due to production of an altered dihydrofolate reductase that has markedly reduced affinity for trimethoprim.

Quinolone Resistance

Resistance to quinolones can be caused by mutations in DNA gyrase subunits A or B, reduced outer membrane permeability in gram-negative cells, or to active efflux transporters found in many bacteria. The highest level of resistance to the newer fluoroquinolones is most frequently associated with chromosomal mutations, causing amino acid substitutions in a highly conserved region in the A subunit of DNA gyrase. Multiple-mechanisms of resistance can occur in a single isolate of bacteria, leading to a higher level of resistance to many fluoroauinolones.

Resistance due to Decreased Entry of a Drug

Tetracycline Resistance

The uptake of tetracycline by members of the Enterobacteriaceae is biphasic. In an initial energy-independent rapid phase, tetracycline binds to cell surface layers and passes by diffusion through the outer layers of the cell. In the second, energy-dependent phase, tetracycline crosses the cytoplasmic membrane, probably by means of a proton-motive force. The precise transport system has not been identified.

Tetracycline resistance is common in both Gram-positive and Gram-negative bacteria. In most cases it is plasmid encoded and inducible; however, chromosomal, constitutive resistance is found in some organisms such as Proteus species. Many plasmid-encoded specified tetracycline resistance determinants have been found in enteric bacteria. The most common of these determinants, TetB, is also present in H influenzae. Tetracycline resistance in Staphylococcus aureus is due primarily to small multicopy plasmids; chromosomal resistance is rare. Tetracycline resistance is found on nonconjugative plasmids in Streptococcus faecalis and on the chromosome of S pneumoniae, S agalactiae (group B streptococci), and oral streptococci. Clostridium species such as C difficile harbor chromosomal genes for tetracycline resistance.

Basically, tetracycline resistance is due to a decrease in the levels of drug accumulation. Decreased uptake and increased efflux both probably participate. Resistant bacteria bind less tetracycline, and the tetracycline they do accumulate is lost by an energy-dependent process when they are in a drug-free milieu.

Plasmid-mediated resistance to tetracyclines can be partially overcome in Gram-positive species by modifying the tetracycline nucleus. Hence, achievable concentrations of minocycline and doxycycline, in particular, will inhibit some tetracycline-resistant streptococci such as S pneumoniae, and some S aureus strains. Molecular modification has not been successful in overcoming the tetracycline resistance of members of the Enterobacteriaceae or Pseudomonas or most Bacteroides species.

Tetracycline resistance is a major concern because it is located on plasmids near insertion sites, and these plasmids readily acquire other genetic information to enlarge the spectrum of resistance. The widespread use of tetracycline in animal feeds may be a factor in the extensive, worldwide resistance of members of the Enterobacteriaceae, particularly enteric species such as Salmonella, to tetracyclines and subsequently to many other drugs. Not only can tetracycline resistance move among members of the Enterobacteriaceae on plasmids, but plasmids mediating tetracycline resistance have moved between S aureus, S epidermidis, S pyogenes, S pneumoniae, and S faecalis.

Fosfomycin Resistance

Fosfomycin and fosmidomycin, which inhibit cell wall synthesis, enter bacteria by means of a glycerol-phosphate or glucose-6-phosphate transport system. Gram-positive bacteria in which the glucose-6-phosphate transport system is poorly developed do not take up these drugs in concentrations adequate to inhibit the cell wall synthesis. This resistance usually is chromosomal. The resistance of Gram-negative bacteria to these agents is related primarily to the presence in the population of some bacteria that can function without the transport system. Plasmids and transposons that transfer resistance to fosfomycin have been found in bacteria such as Serratia marcescens.

Aminoglycoside Resistance

In the most important form of aminoglycoside resistance, the compound is modified outside the cell and resistance is due partly to poor uptake of the altered compound. Also, all aminoglycosides have free amino and hydroxy groups that are essential for binding to ribosomal proteins. A number of enzymes can acetylate the amino groups and phosphorylate or adenylate the hydroxyl groups. Other forms of resistance, such as altered binding site on 30S ribosomes, are much less common.

In members of the Enterobacteriaceae and in Pseudomonas species, the aminoglycosides pass through the cell wall via channels designed to admit cationic molecules to the periplasmic space. These channels, called porin channels, are lined by the porin protein. Aminoglycosides are then translocated across the cell membrane by an energy-dependent proton-motive force and, in the cytoplasm, bind to ribosomes located just below the membrane. Aminoglycosides bind only to ribosomes actively engaged in protein synthesis. Binding to the ribosomes induces a protein involved in the uptake of the aminoglycosides.

Bacteria may contain in the periplasmic space enzymes that acetylate, phosphorylate, or adenylate aminoglycosides to various degrees. It is not clear whether the enzymes are free in the periplasmic space or bound to the cytoplasmic membrane. The modified aminoglycosides do not bind well to ribosomes, and hence uptake is poor or absent.

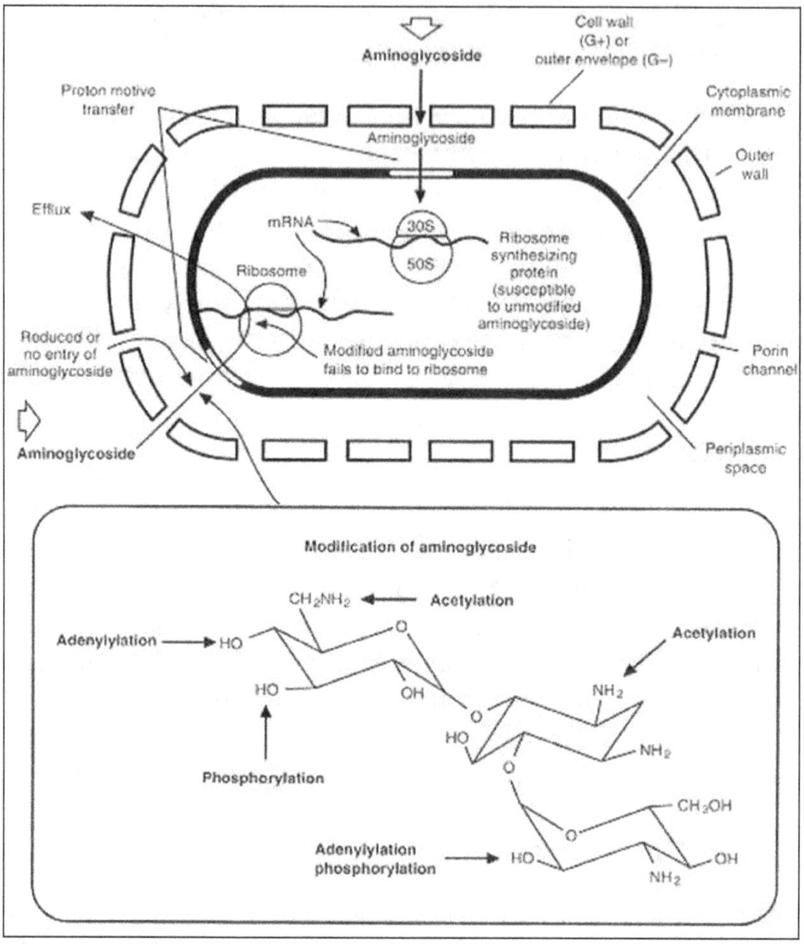

Diagrammatic representation of transfer and transfer reduction of aminoglycoside across the bacterial cell wall. If it is modified by acetylation, adenylation, or phosphorylation, the drug will not bind to ribosomes and will leave the bacterial cell.

Aminoglycoside-modifying enzymes have been found in Gram-positive species such as S aureus, S faecalis, S pyogenes,and S aeruginosa These enzymes are particularly prevalent in members of the Enterobacteriaceae and P aeruginosa. Many of the genes for aminoglycoside-modifying enzymes are carried on transposons.

Anaerobic organisms such as Bacteroides species are resistant to aminoglycosides because they lack an oxygen-dependent transport system to move the drugs across the cytoplasmic membrane. Although most resistance of S aureus to aminoglycosides is due to aminoglycoside-modifying enzymes, small-colony variants of staphylococci also show resistance, which may be due to a defect in adenylate cyclase or in cyclic adenosine 5′-monophosphate (cAMP)binding proteins such that cells with a reduced growth rate do not transport aminoglycosides into the cytoplasm. Some members of the Enterobacteriaceae and P aeruginosa appear to be resistant because of altered porin channels, since these bacteria do not take up any drug and do not have aminoglycoside-inactivating enzymes.

Resistance due to Destruction or Inactivation of a Drug

Chloramphenicol Resistance

Many Gram-positive and Gram-negative bacteria, including some recently discovered H influenzae strains, are resistant to chloramphenicol because they possess the enzyme chloramphenicol transacetylate, which acetylates hydroxyl groups on the chloramphenicol structure. This enzyme, unlike the aminoglycoside-inactivating enzymes and β-lactamases, is an intracellular enzyme of higher molecular weight and subunit structure. Acetylated chloramphenicol binds less well to the 50S ribosome.

ß-Lactam Resistance

The best-known mechanism of bacterial resistance is the resistance to β-lactams, which is mediated by penicillinase enzymes. Resistance of E coli to penicillin was recognized in 1940, before sufficient penicillin was made to be clinically useful. In the 1940s, resistance of staphylococci was shown to be due to a penicillinase. As these enzymes also attack other β-lactam compounds such as cephalosporins, carbapenems, and monobactams, they would be more appropriately designated β-lactamases. The most important activity of these enzymes is alteration of the β-lactam nucleus. β-lactamases are widely distributed in nature and are usually classified on the basis of the principal compounds they destroy (e.g., as penicillinases or cephalosporinases). β-Lactamases may be chromosomally or plasmid mediated, and they may be constitutive or inducible.

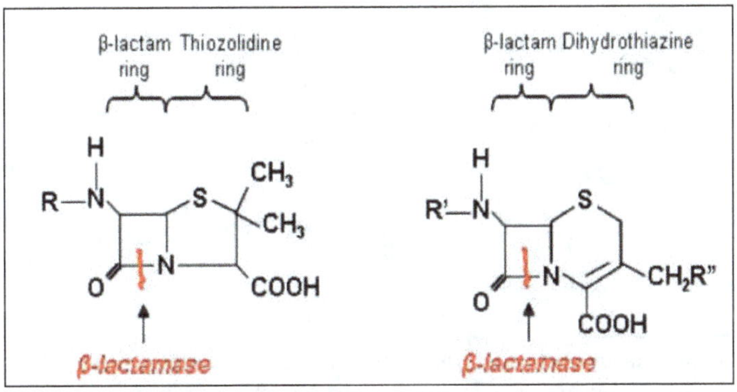

Site of β-lactamase attack in penicillins and cephalosporins.

In Gram-positive species, β-lactamases are primarily exoenzymes; that is, they are excreted into the milieu around the bacteria. Virtually all hospital isolates of staphylococci, both S aureus and S epidermidis, have β-lactamases, and 50 to 80 percent of community-acquired staphylococcal isolates produce β-lactamases. In Gram-negative species, both aerobic and anaerobic, β-lactamases are contained in the periplasmic space, thus effectively protecting the penicillin-binding proteins.

Resistance of staphylococci to β-lactams was soon overcome with the antistaphylococcal penicillins and the cephalosporins. Some strains of S aureus produce more β-lactamase constitutively and can destroy some of the cephalosporins.

In 1974, H influenzae was shown to possess a plasmid mediated β-lactamase. At present 10 to 35 percent of H influenzae strains in the United States produce β-lactamases. The TnA transposon has become more widespread, and the resistance of Haemophilus species to penicillin G and ampicillin seems to be increasing yearly. The Haemophilus β-lactamase is the same structurally as the enzyme found in E coli, Salmonella, Shigella, and N gonorrhoeae. The enzyme has generally been called the TEM enzyme after the initials from whom an E coli strain containing a plasmid β-lactamase was first isolated. These enzymes are also called Richmond-Sykes class IIIa enzymes from a classification proposed by Richmond and Sykes in 1973. By far the most common plasmid β-lactamase found in nature is TEM-1, which accounts for 75 to 80 percent of plasmid-mediated β-lactamase resistance worldwide. Recently new β-lactamases have been found that hydrolyze compounds such as inomethoxy cephalosporin, which were not destroyed by other plasmid-encoded β-lactamases. The new β-lactamases have an altered amino acid composition, which permits binding to the cephalosporin and subsequent hydrolysis. How common these new enzymes will become is unknown.

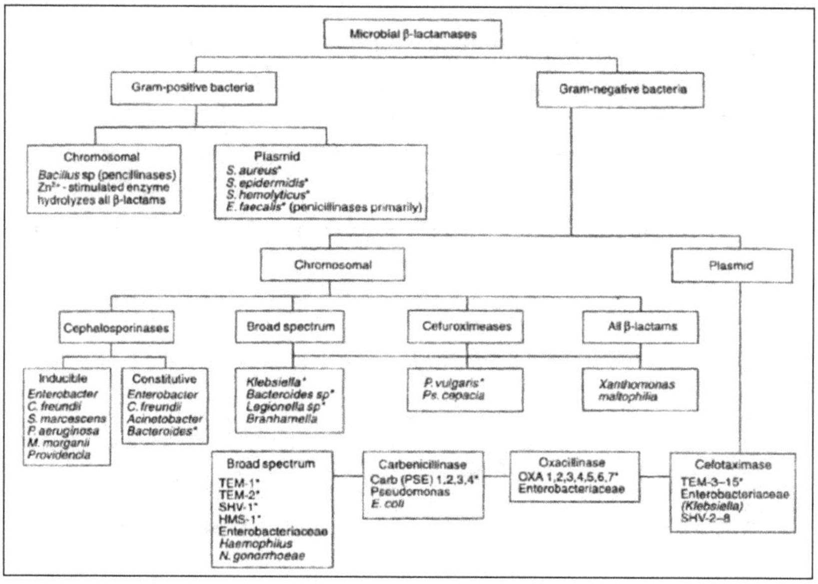

β-Lactamases found in bacteria and their classification and synthesis, whether chromosomally or plasmid mediated.

Chromosomally mediated β-lactamases are present in many Enterobacter, Citrobacter, Proteus-Providencia, and Pseudomonas species. All Klebsiella species possess a β-lactamase, which acts primarily as a penicillinase and is chromosomally mediated. Constitutively produced β-lactamases are also present in many anaerobic species.

Table lists the major β-lactamases of clinical importance. β-Lactamases vary in their ability to destroy penicillins and cephalosporins. β-Lactamase activity is only one component of the β-lactam resistance of Gram-negative bacteria, since resistance to β-lactams is a combination of decreased entry, β-lactamase stability and affinity of the compounds for penicillin-binding proteins.

Synthesis of Resistant Metabolic Pathway

No synthesis of a new type of cell wall resistant to β-lactams has occurred, but some bacteria, particularly some streptococci, lack the hydrolytic enzymes necessary for forming a new cell wall, and so β-lactams do not lyse these bacteria. An altered hydrolytic system thus converts a bactericidal antibiotic into a bacteriostatic agent. Whether such resistance occurs in Gram-negative species is not clear.

Some thymidine-requiring streptococci are not inhibited by trimethoprim and sulfonamides and so are not killed by these agents. These organisms are a rare cause of urinary tract infections. Other bacteria produce adequate deoxyribosylthymine 5′-monophosphate (DTMP) by alternative methods and, as a result. survive exposure to these folate inhibitors.

Certain Candida or Cryptococcus yeasts are resistant to flucytosine because they cannot convert it to its active component, fluorouracil. Other fungi can resist the polyenes and imidazoles because they synthesize membrane components by different metabolic mechanisms.

Combinations of Antimicrobial Agents

Antibiotics are frequently used in combination for the following reasons: (1) to treat a life-threatening infection; (2) to prevent emergence of bacterial resistance; (3) to treat mixed infections of aerobic and anaerobic bacteria; (4) to enhance antibacterial activity (synergy); and (5) to use lower doses of a toxic drug. Combined treatment is reasonable when the precise agents of a serious infection are unknown. Use of two or more drugs to prevent the emergence of resistance is effective for tuberculosis and for therapy of some chronic infections. The use of combinations to achieve synergy is more complicated. Synergy occurs when a combination of two drugs causes inhibition or killing when used at a fourfold-lower concentration than that of either component drug used separately. However, indifference or antagonism may occur instead. Indifference means that the combined action is the same as with either component; antagonism refers to a reduction in the activity of one or both components in the presence of the other.

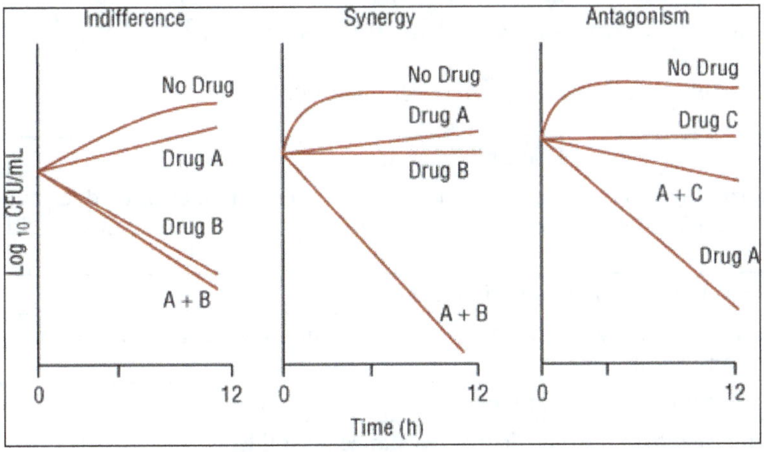

Example of how two antibiotics (A and B) may interact with synergy, indifference, or antagonism.

Important examples of bacterial synergy include (1) combinations of anti-cell wall agents with aminoglycosides, (2) use of β-lactamase inhibitors with β-lactamase-susceptible antibiotics, and (3) combinations of drugs that act on sequential steps in bacterial metabolic or synthetic pathways. Examples of synergy that have proved clinically important are the combination of penicillin and streptomycin to treat Enterococcus faecalis endocarditis and the combination of carbenicillin and gentamicin to treat P aeruginosa infections. Recently the β-lactamase inhibitor clavulanic acid or sulbactam has been combined with amino penicillins to inhibit S aureus, Klebsiella pneumoniae, H influenzae, and anaerobic organisms such as Bacteroides species, all of which are resistant to amoxicillin or ampicillin when they contain β-lactamases. The combination of sulfamethoxazole and trimethoprim attacks two parts of the folic acid cycle and synergistically inhibits many bacteria. Finally, combinations of two penicillins that affect different stages of cell wall synthesis in Gram-negative bacteria are synergistic. This is true for the combination of mecillinam (amdinocillin), which binds to PBP2 of bacteria, and penicillins or cephalosporins, which bind to PBP lb or 3.

Antagonism can occur when a bacteriostatic agent is combined with a bactericidal agent. The classic example has been the combination of chlortetracycline and penicillin in treatment of pneumococcal meningitis. This effect has not been explained from a molecular standpoint, and tetracyclines and penicillins or cephalosporins are used to treat mixed infections such as pelvic inflammatory disease due to N gonorrhoeae and Chlamydia. Some β-lactam antibiotics can induce β-lactamases that inactivate other β-lactams, and antagonism can be shown in the test tube, but the relevance for clinical infections is not established.

Mechanism to Reduce Bacterial Resistance

Proper selection of new antibiotics will be a major force in slowing the development of antimicrobial resistance. Proper hygiene practices will reduce plasmid transfer and the establishment of multiple drug-resistant bacteria in the hospital and will delay the appearance of such species in the community. Table lists a number of mechanisms to prevent bacterial resistance. The health care provider must be continually alert to the appearance of antibiotic resistance within the hospital and community.

Table: Mechanisms to reduce antibiotic resistance.

1.	Control reduce or cycle antibiotic usage.
2.	Improve hygiene in hospitals and among hospital personal and reduce movement of patients to eliminate the dissemination of resistant organisms within hospitals.
3.	Discover or develop new antibiotics.
4.	Modify existing antibiotics chemically to produce compounds to known mechanisms of resistance.
5.	Develop inhibitors of antibiotic-modifying enzymes.
6.	Define agents that would "cure" resistance plasmids.

DURATION OF ANTIMICROBIAL THERAPY

Deciding on the duration of antimicrobial therapy for aon infection is neither straightforward nor simple. A short or long course of antibiotics can be given to a patient, depending on the drug used, the severity of an infection, and response to treatment.

Although antibiotics are, in general, safe, they also have many risks associated with their use, including the development of allergic reactions, Clostridium difficile infection, and antibiotic resistance, as well as a higher price tag. As such, many clinicians prefer prescribing a shorter treatment course. However, it also is important to provide a substantial treatment course so that an infection is treated adequately and relapse is prevented.

Choice of Agent

Antibiotics work by either selectively killing (bactericidal) or inhibiting the growth (bacteriostatic) of bacteria. Infections with a high bacterial burden, such as those seen in infective endocarditis, require treatment with antibiotics with rapid bactericidal activity. In most cases, more than 1 antibiotic is used to provide synergistic activity and rapid killing. For example, when treating native valve endocarditis caused by penicillin-susceptible viridans streptococci, the usual treatment duration can vary from 2 to 4 weeks. If penicillin G or ceftriaxone monotherapy is prescribed, the treatment duration should be 4 weeks; however, if either antibiotic is used in combination with an aminoglycoside, then the treatment can be shortened by 2 weeks.

Another factor that can affect the efficacy of antibiotics is their ability to penetrate and remain (for an adequate time) at the site of infection. Fosfomycin tromethamine, quinolones, nitrofurantoin, trimethoprim-sulfamethoxazole and beta-lactams are some of the antibiotics used to treat urinary tract infections. Even though these antibiotics can concentrate well in the genitourinary tract, each can differ in duration of treatment. For example, fosfomycin can be given in a 1-time dose because a single 3-g oral dose can provide a peak urinary concentration within 4 hours and remain elevated (>128 mg/L) for ≤48 hours. In another example, Hooton et al found that a 3-day course of cefpodoxime proxetil, a third-generation cephalosporin, did not provide a similar cure rate as it did with a 3-day regimen of ciprofloxacin (cure rate, 82% vs 93%, respectively; 95% CI, 3%-18%) for the treatment of acute uncomplicated cystitis.

Type and Severity of Infection

A patient's immune status, affected anatomical site, and pathogen causing the infection also should be considered when assessing duration of therapy. Infected patients with no hemodynamic issues can be treated with short courses of antimicrobials. For example, community-acquired pneumonia (CAP) can be treated in as little as 5 days, but once the patient's condition is complicated by bacteremia or severe sepsis, a longer course of antibiotics is essential.

The ability of antibiotics to penetrate necrotic tissues, abscesses, or biofilms also can limit their efficacy. Infections can be difficult to treat and require prolonged antibiotic courses. Unless surgical intervention is undertaken to remove debris and/ or drain abscesses, antibiotics cannot reach infected sites. For example, a course of antibiotics for intraabdominal infections is no longer than 7 days; however, if it is difficult to perform the source control procedure (eg, drain infected foci, control ongoing peritoneal contamination), a longer treatment course is necessary.

Assessment of Response

Improvements in hemodynamic status (eg, heart rate, blood pressure), white blood cell count, temperature, oxygenation, and/or radiologic findings should be seen a few days after starting an effective therapy. Once the signs and symptoms of infections are resolved, clinicians can consider terminating therapy. El Moussaoui et al conducted a randomized, double-blind, placebo-controlled study comparing the effectiveness of discontinuing amoxicillin therapy in mild to moderate-severe CAP after 3 days compared with 8 days. Patients were assessed regarding respiratory symptoms (dyspnea, cough, sputum production, color of sputum) and general improvement (not recovered to complete recovery) based on a 5-point symptom scale. Patients who improved by ≥2 points, who had a temperature <38°C, and who were able to take oral medications were randomized to receive either 750-mg oral amoxicillin or placebo for 3 to 5 days. They found that discontinuing antibiotics 3 days after symptom resolution did not adversely affect patient outcomes. In addition, there were no differences in clinical or radiological outcomes between the 2 groups after 10 days and 28 days.

Rechecking for cultures is not always necessary once a patient begins responding to therapy, except in the case of bloodstream infections. Monitoring for bacterial clearance is crucial because day 1 of antimicrobial therapy is the first day on which negative blood cultures are obtained.15 Acquiring unnecessary cultures should be avoided because a positive culture having no signs and symptoms of infection could lead to treating colonized bacteria.

The use of biomarkers, such as C-reactive protein (CRP), and the procalcitonin test also has been instrumental in evaluating antibiotic response and determining the duration of antibiotic therapy. Unlike CRP, procalcitonin is more specific to bacterial infections; therefore, the test has been used to curtail unnecessary antibiotic usage. Use of the procalcitonin-guided algorithm has been shown to reduce the duration of exposure to antibiotics by ≤25% in patients with lower respiratory tract infections and 23% in patients who are critically ill.

COMBINATION THERAPY

Combination antibiotic therapy for invasive infections with Gram-negative bacteria is employed in many health care facilities, especially for certain subgroups of patients, including those with neutropenia, those with infections caused by Pseudomonas aeruginosa, those with ventilator-associated pneumonia, and the severely ill. An argument can be made for empiric combination therapy, as we are witnessing a rise in infections caused by multidrug-resistant Gram-negative organisms. The wisdom of continued combination therapy after an organism is isolated and antimicrobial susceptibility data are known, however, is more controversial. The available evidence suggests that the greatest benefit of combination antibiotic therapy stems from the increased likelihood of choosing an effective agent during empiric therapy, rather than exploitation of in vitro synergy or the prevention of resistance during definitive treatment.

Multidrug-resistant Gram-negative organisms (MDRGNs) have emerged as a major threat to hospitalized patients and have been associated with mortality rates ranging from 30 to 70%. The abundant and often inappropriate use of broad-spectrum antibiotics contributes to the emergence of MDRGNs. A vicious cycle is created as MDRGN infections force us to rely on additional broad-spectrum antibiotics to treat these infections, leading to yet more resistance. The emergence and proliferation of these highly resistant Gram-negative organisms are particularly concerning given the limited number of antimicrobial agents that are currently available or in the drug development pipelines of the pharmaceutical industry to combat these organisms. A reduction in inappropriate utilization of broad-spectrum antibiotics is clearly important to minimize the emergence of MDRGNs. Every effort needs to be made to carefully select antibiotics, balancing the need for a broad spectrum of empiric coverage of potential microorganisms with the need to preserve available antibiotics for when they are absolutely necessary.

One area where the approach to antibiotic use needs to be readdressed is the use of combination antibiotic therapy, which generally consists of a β-lactam and an aminoglycoside or fluoroquinolone, for the treatment of infections with Gram-negative bacteria. There is evidence supporting the initial use of combination therapy for severe infections with Gram-negative bacteria, such as sepsis or ventilator-associated pneumonia (VAP), in the existing environment of MDRGNs because of the broad empiric coverage provided by two antimicrobial agents with different spectra of activity. However, when identification and susceptibility testing results are known, an argument can be made that the antibiotic regimen for Gram-negative organisms can be "fine-tuned" and narrowed in many cases.

Observational studies show that between 25 and 50% of patients with bacteremia, surgical site infections, or pneumonia and over 50% of patients with septic shock in the intensive care unit (ICU) are administered combination antibiotic therapy. The question of whether a combination of a β-lactam and an aminoglycoside or fluoroquinolone confers a benefit in patients beyond broadening the antimicrobial spectrum during the empiric treatment period before culture results are available is unsettled. With the availability of new broad-spectrum and highly bactericidal antibiotics, the need to combine β-lactams with a second agent for the treatment of infections with Gram-negative bacteria should be reassessed.

Intuitive Appeal of Combination Therapy

Whether combination antimicrobial therapy is more efficacious than monotherapy for infections with Gram-negative bacteria remains controversial, particularly for infections due to organisms more commonly acquired in hospital settings, such as Pseudomonas spp., Serratia spp., Acinetobacter spp., and Enterobacter spp. Traditionally, combination antibiotic therapy for infections with Gram-negative bacteria has included two agents to which an organism demonstrates in vitro susceptibility, typically a β-lactam and an aminoglycoside. Although there are theoretical advantages to combination therapy shown by in vitro and animal studies, clinical data have been conflicting.

The initial use of combination therapy for infections with Gram-negative bacteria is often justified by one of the following three reasons: (i) to broaden the empiric coverage provided by two antimicrobial agents with different spectra of activity (an effort to ensure that the pathogen is adequately covered by at least one of the two components of the regimen), (ii) to exploit the synergy observed in vitro between two antibiotic agents compared to one (and hence improve clinical outcomes), or (iii) to prevent or delay the emergence of resistance during antimicrobial therapy.

Despite the intuitive appeal of these approaches, strong evidence supporting the use of two antimicrobials to treat infections with Gram-negative bacteria is lacking, and there is evidence that it may even be harmful. The addition of a second antimicrobial agent to treat a Gram-negative organism that is susceptible to a single agent may actually lead to increased antimicrobial resistance, adverse effects, and costs. The development of adverse effects such as aminoglycoside-related nephrotoxicity is well documented. Some may argue, however, that definitive combination therapy (as opposed to empiric therapy) may still be warranted for certain subpopulations or circumstances. For example, some have suggested that there is an advantage to the prescription of combination therapy for profoundly neutropenic patients, patients with Pseudomonas aeruginosa sepsis, intensive care unit (ICU) patients, patients with VAP, or septic patients with significantly elevated severity-of-illness scores. The evidence supporting or refuting these claims is detailed below.

Arguments in Favor of Combination Therapy

Broad Spectrum of Activity

In the age of increasingly resistant Gram-negative infections, the likelihood that empiric antimicrobial therapy will provide adequate coverage for potential pathogens causing an infection is increased with the use of two antimicrobial agents compared to a single agent. Prompt institution of antimicrobial therapy active against the causative pathogen is crucial in the treatment of severely ill patients suspected of having a bacterial infection. The use of at least one antimicrobial agent to which a pathogen is susceptible for empiric therapy leads to lower mortality and improved outcomes in patients with sepsis caused by Gram-negative bacteria, as observed in a number of studies. Mortality rates are higher among patients with health care-associated infections when they are initially treated with an empiric antimicrobial agent lacking in vitro activity against the infecting pathogen. Evidence exists that patients infected with MDR organisms are more likely to experience a delay in the initiation of effective antimicrobial therapy, and some of this risk can be avoided with the addition of a second agent.

Studies demonstrating a benefit with empiric combination therapy. Kumar et al. conducted a retrospective, propensity-matched cohort study involving 28 ICUs to evaluate the therapeutic benefit of empiric combination therapy (β-lactams in combination with aminoglycosides, fluoroquinolones, or macrolides/clindamycin) compared with β-lactam monotherapy in 4,662 eligible cases of culture-positive bacterial septic shock. Empiric combination therapy was associated with a decreased 28-day mortality (36% versus 29%; P = 0.0002) and increases in both mechanical ventilation-free days (median [interquartile range], 10 [0 to 25] versus 17 [0 to 26]; P = 0.008) and pressor-free days (23 [0 to 28] versus 25[0 to 28]; P = 0.007). Notably, antipseudomonal penicillins, antipseudomonal cephalosporins, and carbapenems failed to exhibit a benefit with the addition of a second agent. This may be due to the broad spectra of activity of these agents against the vast majority of Gram-negative pathogens responsible for septic shock, with minimal incremental benefit from the addition of a second agent. The rationale for inclusion of clindamycin/macrolides alone or in combination with other agents is unclear, as these agents are not routinely prescribed for septic shock, with a notable exception being clindamycin for toxin-mediated shock. clindamycin or macrolides were used in approximately 15% of patients either as single agents or as "combination" therapy.

Similarly, a retrospective cohort analysis establishing the relationship between initial inappropriate antimicrobial treatment and the clinical outcomes for P. aeruginosa infections showed that hospital mortality was significantly higher for patients receiving inappropriate initial antimicrobial treatment than for those receiving appropriate therapy (31% versus 18%; P < 0.02). Inappropriate therapy was defined as "the absence of Gram-negative antimicrobial agents with in vitro activity against P. aeruginosa." Inappropriate initial administration occurred more frequently among patients receiving monotherapy (35% versus 21%; P = 0.01). The All-Patient Refined Diagnosis-Related Group (APR-DRG) score was used to measure severity of illness. Patients receiving inappropriate initial antimicrobial treatment had statistically greater APR-DRG scores (P = 0.01); however, APR-DRG scores were not incorporated into the final models, making it difficult to determine if mortality was attributable to empiric antibiotic choices or to the underlying severity of illness.

Similar results were achieved in other studies, with mortality in excess of 30% and an increased length of hospital stay related to delays in the initiation of appropriate therapy in ICU patients with sepsis caused by Gram-negative bacteria. These studies suggest that inappropriate antimicrobial treatment can be reduced with empiric administration of combination therapy. One must recognize, however, that it is difficult to adequately control for illness severity in these studies, and therefore it is difficult to assess the excess attributable mortality due to inadequate empiric antimicrobial therapy. When empiric combination therapy is prescribed, the second agent that is selected should have activity against an organism potentially resistant to the β-lactam agent.

Use of local antimicrobial epidemiology to inform empiric antibiotic choices. The selection of empiric combination therapy for presumed infections with Gram-negative bacteria needs to be made after considering local epidemiology and individual patient characteristics. Prior to prescribing antimicrobial therapy, resistance patterns within an institution are important to consider, and close liaison with the microbiology laboratory facilitates the decision-making process. Surveillance data and hospital-specific antibiograms inform empiric antibiotic choices. Data from U.S. ICU studies, surveillance studies, the National Nosocomial Surveillance System, and the SENTRY Antimicrobial Surveillance Program have shown that overall susceptibilities within a 10-year period have declined significantly for all drug classes studied, and these trends are likely mirrored in individual institution antibiograms as well. Nationally, for example, multidrug resistance of P. aeruginosa to three or more antipseudomonal agents rose from 4% in 1993 to 14% in 2002 ($P < 0.001$), and it continues to rise.

Increasing Gram-negative resistance complicates the selection of empiric therapy in severe infections. This is highlighted in a study by Lautenbach et al., who found that the time to effective therapy for infections due to extended-spectrum β-lactamase (ESBL)-producing Gram-negative bacilli was approximately six times longer than that for infections caused by non-ESBL-producing strains (medians, 72 h versus 11.5 h). These results are supported by a meta-analysis of 16 studies which found that ESBL production is associated with a delay in effective antimicrobial therapy for patients with Gram-negative bacteremia and a subsequent increased mortality.

When resistance to β-lactam therapy is anticipated in patients with sepsis presumed to be caused by Gram-negative bacteria, the addition of an aminoglycoside until antimicrobial susceptibilities are known appears to be justified. It is important to evaluate local antibiograms to determine which aminoglycoside would be most likely to increase the range of coverage against Gram-negative bacteria. However, when the β-lactam agent is sufficiently broad (e.g., carbapenem) and there is no local epidemiologic evidence supporting the likelihood of highly resistant organisms, the benefit of combination therapy, even empirically, is unclear.

Individualization of empiric therapy based on patient characteristics.Information on local antimicrobial resistance patterns should be supplemented with patient-specific characteristics to guide empiric treatment choices. Individualization of initial empiric antibiotic therapy is essential, as patients differ with regard to preexisting medical conditions, severity of illness, nature of infection, previous antibiotic and hospital exposure, presence of indwelling catheters, and colonization with antibiotic-resistant organisms.

Bhat et al. demonstrated how knowledge of recent receipt of antibiotics and colonizing flora can improve the adequacy of initial empirical therapy. Thirty-seven percent of patients receiving piperacillin-tazobactam in the month prior to their current infection were infected with piperacillin-tazobactam-resistant P. aeruginosa in the subsequent month. Piperacillin-tazobactam was considered to be appropriate empiric therapy only if during the prior month, the patient had neither received the antibiotic nor had isolation of a piperacillin-tazobactam-resistant organism. In these situations, broadening empiric therapy to a carbapenem or addition of an aminoglycoside improved the likelihood of adequate empiric coverage.

Deescalation of antimicrobial therapy when susceptibility results are known. Although ample evidence demonstrates that initial prescription of combination therapy may be beneficial in a septic patient potentially infected with an MDRGN, when identification and susceptibility testing are complete, the antibiotic regimen should be "fine-tuned." Narrowing antimicrobial therapy based on antibiotic susceptibility results and appropriately limiting the duration of therapy are the cornerstones of responsible antimicrobial prescription.

Synergy

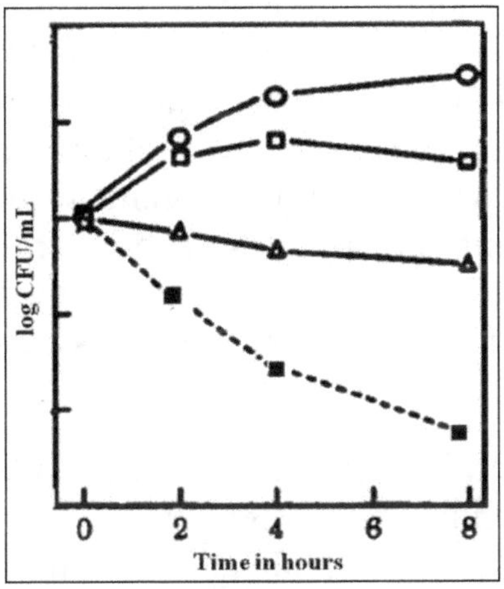

Time-kill kinetics demonstrating growth of organisms in the settings of no drug (circles), addition of drug A (open squares), addition of drug B (triangles), and addition of both drugs A and B (closed squares).

A potential benefit of adding a second antimicrobial agent is the synergistic effect of the combination (i.e., more rapid killing of the pathogen). Synergy between two antimicrobial agents is defined as a greater-than-2-log increase in bactericidal activity in vitro compared with the bactericidal activity of each agent alone. The rate of bacterial killing by a fixed concentration of a single agent or multiple agents in combination can be depicted by a time-kill curve. Alternatively, various concentrations of two different agents can be used to evaluate their synergistic effect using the checkerboard technique.

β-Lactam and aminoglycoside synergy in in vitro and animal models. For infections with Gram-negative bacteria, antimicrobial synergy has traditionally been seen with β-lactam–aminoglycoside combinations. The combination of a β-lactam and an aminoglycoside allows for different mechanisms of bacterial killing. β-Lactam-mediated disturbance of the cell walls of Gram-negative bacilli facilitates passage of aminoglycosides into the periplasmic space.

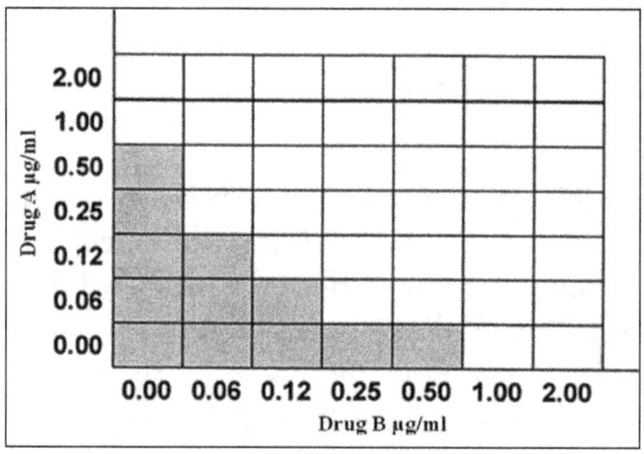

Synergy of a two-drug combination determined using the checkerboard technique.

Synergy was initially studied in enterococcal endocarditis in in vitro models. These studies demonstrated that penicillin enhanced the uptake of aminoglycosides. Other in vitro studies, albeit using small numbers of isolates, have similarly found that treatment with a combination of a β-lactam and an aminoglycoside is superior to treatment with a β-lactam alone. Synergy was not observed in an in vitro study of P. aeruginosa when a dense inoculum of microorganisms was present. In vitro synergy appears to be variably present and strain dependent and varies with different β-lactam and aminoglycoside combinations (109). With the use of newer, broader-spectrum β-lactams, it is unclear if the results of these studies would be different if the studies were repeated.

A viridians group streptococcal rabbit endocarditis model was used to determine in vivo synergism between penicillin and streptomycin. The use of low-dose gentamicin in rabbits experimentally infected with Staphylococcus aureus demonstrated more rapid bactericidal activity of S. aureus from cardiac vegetations when initial low-dose gentamicin was combined with antistaphylococcal penicillins than when antistaphylococcal penicillins were used alone. With combination therapy, the bacteria were eradicated from the cardiac vegetations in half the time required to achieve the same results with penicillin monotherapy. Similar findings were seen in Gram-negative animal models. Rabbits with P. aeruginosa endocarditis treated with 2 weeks of combination therapy with carbenicillin and gentamicin were significantly more likely to have sterilization of vegetations than rabbits treated with 2 weeks of carbenicillin therapy alone.

β-Lactam and aminoglycoside synergy. While synergistic action between β-lactams and aminoglycosides has been shown in vitro, clinical evidence to support these data are sparse and conflicting. A prospective cohort study of 200 patients with P. aeruginosa bacteremia (both neutropenic and nonneutropenic patients) was undertaken to compare in vitro susceptibility results with mortality. No significant correlation between in vitro synergy testing (either time-kill or checkerboard) and clinical outcome was demonstrated. Additionally, results obtained by time-kill curve and checkerboard synergistic testing were not correlated; combination therapy found to be synergistic by one method was not necessarily synergistic by the other method. A similar study compared in vitro synergy testing with clinical outcomes in 14 nonneutropenic patients infected with P. aeruginosa. Clinical cure was defined as the resolution of symptoms and signs of infection. The investigators found no clinical evidence of increased likelihood of clinical cure in patients treated with a β-lactam and an aminoglycoside, regardless of in vitro testing.

In contrast, in a retrospective study of 444 cases of Gram-negative bacteremia, there was an 80% clinical response rate in patients who received antibiotic therapy that was synergistic against the organism (using the checkerboard technique), compared to a 64% response rate in patients who received nonsynergistic combinations ($P < 0.05$). Synergism in vitro was correlated with better clinical responses in patients with neutropenia, shock, and P. aeruginosa infections.

In a second retrospective study of profoundly neutropenic patients with Gram-negative bacteremia, a clinical response was observed in 7 of 11 (64%) of patients in whom synergism was present (as defined by the checkerboard technique) compared to 0 of 6 patients when synergism was not present (0%). The authors concluded that synergistic combinations were indicated for profoundly neutropenic patients with Gram-negative bacteremia.

There are discrepancies when comparing in vitro and in vivo studies assessing combination therapy for infections with Gram-negative bacteria. When weighing the unclear benefit of in vivo synergy with the potential negative consequences of combination therapy (i.e., nephrotoxicity, ototoxicity, additional monitoring requirements, etc.), the rationale for combination therapy becomes questionable.

β-Lactam and fluoroquinolone synergy in in vitro models. Although antimicrobial synergy appears to be best established for β-lactam–aminoglycoside combinations, similar data on synergistic activity have emerged for combinations of β-lactams and fluoroquinolones. In vitro synergy between β-lactams and fluoroquinolones against Gram-negative organisms has ranged from 17% to 82%. One study evaluated ciprofloxacin in combination with imipenem versus ciprofloxacin and amikacin against clinical isolates of multidrug-resistant P. aeruginosa; 42% (11/26) of strains demonstrated synergy with the combination of ciprofloxacin and imipenem, whereas only 15% (4/26) of isolates demonstrated enhanced killing with the combination of ciprofloxacin and amikacin. An in vitro study of 12 clinical isolates of P. aeruginosa found no difference in the degree of synergy between β-lactam–aminoglycoside and β-lactam–fluoroquinolone combinations, with synergy percentages ranging from 58% to 79%.

In an in vitro study assessing synergy against Burkholderia cepacia, ciprofloxacin in combination with imipenem demonstrated synergy against 44% (7/16) of isolates. Ciprofloxacin in combination with ceftazidime, aztreonam, or azlocillin produced synergy rates of ≥50% against 108 isolates of P. aeruginosa resistant to ciprofloxacin but susceptible to the β-lactam used in the combination in another in vitro study using the checkerboard technique. However, in cases where the P. aeruginosa isolates were susceptible to both antibiotics, the synergy was <20%; if the isolates were resistant to the β-lactam but susceptible to ciprofloxacin, the synergy rate was <5%. Pohlman et al. evaluated ciprofloxacin synergy in combination with aztreonam, ceftazidime, piperacillin-tazobactam, and ticarcillin-clavulanic acid against various Gram-negative organisms. They concluded that synergy between ciprofloxacin and β-lactams was sporadic and was not consistent across drug concentrations or sampling times. Existing data do not demonstrate synergistic activity between fluoroquinolones and aminoglycosides. The synergistic potential of β-lactams and fluoroquinolones remains unclear.

β-Lactam and fluoroquinolone synergy. Fluoroquinolones are recognized for excellent tissue penetration into lung, meninges, and bone, and they have minimal nephrotoxicity compared with aminoglycosides. Al-Hasan et al. conducted a retrospective cohort study incorporating propensity scores evaluating 28-day mortality in 702 patients with Gram-negative bacteremia receiving a combination of β-lactam and fluoroquinolone or β-lactam monotherapy. Combination therapy was associated with lower 28-day mortality than monotherapy (4.2% versus 8.8%; adjusted hazard ratio [HR], 0.44; 95% confidence interval [CI], 0.20 to 0.98; P = 0.04); however, the additional benefit of fluoroquinolones was not evident for critically ill patients. The authors believed that increased mortality in severely ill patients, regardless of the use of combination therapy, may have been the result of other patient factors, including multiorgan failure, systemic inflammatory response, and other underlying medical conditions.

A meta-analysis of 8 randomized, controlled trials (RCTs) was conducted, comparing a β-lactam and ciprofloxacin to a β-lactam and aminoglycoside for the treatment of patients with febrile neutropenia. Clinical cures in the subset of patients with documented infections (odds ratio [OR], 1.56; 95% CI, 1.05 to 2.31) and mortality (OR, 0.85; 95% CI, 0.54 to 1.35) were no different between the two groups; however, nephrotoxicity (OR, 0.30; 95% CI, 0.16 to 0.59) was notably decreased in patients receiving combination therapy incorporating fluoroquinolones.

According to the available evidence, in vitro synergy does not necessarily translate into a clinical benefit. In vitro synergy studies are conducted in well-controlled environments where precise concentrations of multiple antibiotics are tested against known inocula of microorganisms, which can be very different from the unpredictable drug concentrations and microorganism burdens of actual patients. Additionally, in vitro studies cannot take into account the added contribution of the host immune system.

Synergy testing in the microbiology laboratory. The use of combination antibiotic susceptibility testing to guide clinical decisions is generally limited to multidrug-resistant organisms in the cystic fibrosis population. Previous studies suggest that this method of testing yields reproducible results. However, clinical studies have not consistently correlated in vitro synergy results with improved clinical outcomes, and thus such testing is performed on a very limited basis. Often, by the time synergy testing on an isolate is complete, the resistance profile for the organism infecting a patient has already changed. Additionally, synergy testing is generally performed on planktonically growing bacteria (i.e., free-floating bacteria) as opposed to bacteria in biofilms. Bacteria growing in biofilms are generally significantly more resistant, and biofilms often coat the airways of patients with cystic fibrosis Lastly, synergy testing is performed on the sickest patients with highly resistant organisms who have a very poor prognosis, and synergistic combinations may provide minimal improvements in clinical outcomes.

Prevention of Resistance

The available evidence shows that the proportion of Gram-negative organisms resistant to commonly used antibiotics is increasing. Resistance is even observed with antibiotics considered "salvage" therapy, such as tigecycline and colistin. Unfortunately, as antimicrobial resistance is worsening, the antimicrobial armamentarium against Gram-negative bacilli remains relatively constant. Carmeli et al. published one of the first studies to address outcomes associated with antimicrobial resistance in Gram-negative pathogens. The emergence of resistance was associated with a 3-fold-greater risk of death ($P = 0.02$) and a 1.7-fold-longer duration of hospital stay ($P < 0.001$). The estimated mean adjusted increase in duration of hospitalization was 5.7 days. Lautenbach et al. found higher in-hospital mortality among patients infected or colonized with aztreonam-resistant or fluoroquinolone-resistant P. aeruginosa strains than among patients with more susceptible P. aeruginosa strains.

One possible rationale for the use of combination antimicrobial therapy is to prevent or delay the emergence of resistance during treatment. However, convincing clinical data supporting this theory are lacking.

In vitro evidence for prevention of resistance. A notable example of successful use of combination antimicrobial therapy for prevention of resistance is the treatment of infections due to Mycobacterium tuberculosis. Combination therapy for M. tuberculosis infections significantly decreased the rate of development of resistance to rifampin. However, with the relatively slower growth of M. tuberculosis and slower emergence of resistance mutations, combination therapy may not yield comparable results for infections with Gram-negative bacteria. An in vitro study of the combination of azlocillin and tobramycin was undertaken to assess the development of resistance in eight P. aeruginosa isolates recovered from patients with cystic fibrosis after 12 to 16 treatments. Upon exposure to azlocillin and tobramycin combinations using the checkerboard method, the MICs of neither azlocillin nor tobramycin changed significantly; however, there was some evidence of an increase in MICs when the individual antibiotics were used alone. Similarly, in two small in vitro studies of five strains and three strains of P. aeruginosa, respectively, the combination of levofloxacin and imipenem appeared to delay the emergence of resistance.

Clinical data supporting combination therapy for prevention of resistance. In general, most clinical studies comparing combination therapy and monotherapy focus primarily on clinical effectiveness and toxicities and are generally not designed to study the emergence of antimicrobial resistance as a primary outcome. Most clinical studies have had inadequate samples sizes to make conclusions regarding emergence of resistance when comparing combination therapy to monotherapy.

Gribble et al. compared piperacillin as a single agent with carboxypenicillin-aminoglycoside combinations in a prospective, randomized trial of 50 adults with serious bacterial infections. The difference in the clinical response rates between the two regimens was not significant; however, the emergence of resistant organisms during therapy was more frequent among patients receiving piperacillin alone (42%) than among patients receiving combination therapy (17%) ($P < 0.05$). The authors concluded that the use of piperacillin as a single agent in the treatment of serious bacterial infections should not be advocated.

The incidence of emergence of resistance was also evaluated by retrospective review of 173 studies encompassing over 14,000 patients. Bacterial resistance occurred among 5.6% of infections and appeared to be significantly more frequent with penicillin and aminoglycoside monotherapy. Lower rates were associated with broader antimicrobial combinations such as carbapenems and combination therapy. Of importance, aminoglycoside therapy used alone can be an independent risk factor for resistance, and it is unknown if these results would have been replicated if monotherapy was limited to agents other than aminoglycoside.

4

Multiple Drug Resistance

The antimicrobial resistance which is shown by a species of microorganisms to multiple drugs is referred to as multiple drug resistance. Multidrug tolerance and resistance in bacteria and fungi are studied within it. The topics elaborated in this chapter will help in gaining a better perspective about the subject of multiple drug resistance.

Multidrug resistance (MDR) is defined as insensitivity or resistance of a microorganism to the administered antimicrobial medicines (which are structurally unrelated and have different molecular targets) despite earlier sensitivity to it. According to WHO, these resistant microorganisms (like bacteria, fungi, viruses, and parasites) are able to combat attack by antimicrobial drugs, which leads to ineffective treatment resulting in persistence and spreading of infections. Although the development of MDR is a natural phenomenon, extensive rise in the number of immunocompromised conditions, like HIV-infection, diabetic patients, individuals who have undergone organ transplantation, and severe burn patients, makes the body an easy target for hospital acquired infectious diseases, thereby contributing to further spread of MDR. Studies from WHO report have shown very high rates of resistance in bacteria such as Escherichia coli against antibiotics as cephalosporin and fluoroquinolones, Klebsiella pneumoniae against cephalosporin and carbapenems, Staphylococcus aureus against methicillin, Streptococcus pneumoniae against penicillin, Nontyphoidal Salmonella against fluoroquinolones, Shigella species against fluoroquinolones, Neisseria gonorrhoeae against cephalosporin, and Mycobacterium tuberculosis against rifampicin, isoniazid, and fluoroquinolone causing common infections(like urinary tract infections, pneumonia, and bloodstream infections) and high percentage of hospital-acquired infections. A limited number of antifungal drugs are available for the treatment of chronic fungal infections. Resistance to drugs such as polyene macrolides (amphotericin B), azole derivatives (ketoconazole, fluconazole, itraconazole, and voriconazole), DNA and RNA synthesis inhibitors (flucytosine), and 1,3-β-glucan synthase inhibitors (echinocandins) exists in isolates of Candida spp., Aspergillus spp., Cryptococcus neoformans, Trichosporon beigelii, Scopulariopsis spp., and Pseudallescheria boydii. Prolonged drug exposure and nonstop viral replication result in the advent of various resistant strains and persistence of infections despite therapy. This has made antiviral resistance a matter of concern in immunocompromised patients. Consequences of antiviral drug resistance were observed in immunosuppressed transplant recipients and oncology patients infected by either cytomegalovirus (CMV), herpes simplex virus (HSV), Varicella-zoster virus (VZV), human immunodeficiency virus (HIV), influenza A virus, hepatitis C (HCV), or hepatitis B virus (HBV). Parasitic multidrug resistance has been analyzed in isolates of Plasmodia, Leishmania, Entamoeba, Trichomonas vaginalis, schistosomes, and Toxoplasma gondii against drugs such as, chloroquine, pyrimethamine, artemisinin, pentavalent antimonials, miltefosine, paromomycin, and amphotericin B as well as atovaquone and sulfadiazine. One of the most prime examples of disease prone to MDR is malaria, caused by Plasmodium falciparum. Another protozoan parasite, Entamoeba spp., causes amoebiasis which is also a major public health threat in many tropical and subtropical countries. A global health threat of schistosomiasis is also considered similar to that of malaria and other chronic diseases.

Significance of MDR

Antimicrobial drugs have been used for several decades across the world. Surveillance in different regions of the world such as Africa, some parts of America, Eastern Mediterranean Region, Europe, South-East Asia, and Western Pacific Region has shown that many infectious microorganisms have evolved over the years and there is an alarming high number of antibiotic-resistant species enabling themselves to resist the inhibitory effects of these drugs. Not only a single but almost all the capable infecting agents (e.g., bacteria, fungi, virus, and parasite) have employed high levels of MDR with enhanced morbidity and mortality and, thus, are referred to as "super bugs." Tuberculosis, pneumonia, HIV, influenza, malaria, yeast infections, and many other deadly diseases are major causes of deaths in modern era, therefore, indicating MDR as a serious worldwide threat to public health. The chances of controlling tuberculosis have decreased due to resistance of MTB to respective antibiotics, thus, making it a global concern. An overall 6% of recent TB cases and 20% of formerly treated TB cases are likely to possess MDR, while 92 countries were found to have extensively drug resistant TB (XDR-TB). Another bacterial infection, pneumonia, has become untreatable because its causative agent has been found to be resistant to cephalosporin as well as carbapenems due to extended spectrum β-lactamases (ESBL) mediated mechanism , thereby rendering all available treatment using β-lactam antibiotics. In recent years, HIV drug resistance has driven the antiretroviral therapy failure to such an extent that it is charging exorbitant rates along with a number of side effects. The protozoan parasite responsible for malaria had embarked on showing resistance to some of its most effective drugs, chloroquine, artemisinin, and pyrimethamine. This has resulted in replacement of these old ineffective drugs by novel drugs, which has increased the health care expenses. The emergence of resistance to antifungal drugs in invasive yeast infections, for example, Candidiasis, has led to worldwide morbidity and mortality, contributing to global economic burden. Antimicrobial resistance (AMR) or MDR is the reason why microbes fail to respond to standard drugs, thus, extending the duration of course of treatment further increasing the health care costs which tend to worsen the situation of people who are not capable of such expenses.

Problems Associated with MDR

Antimicrobial resistance is associated with high mortality rates and high medical costs and has a significant impact on the effectiveness of antimicrobial agents. MDR provokes obstruction in disease control by intensifying the possibility of spreading of resistant pathogens, thus, declining efficacy of treatment and, hence, resulting in prolonged time of infection in patient. The cost of treatment is also increased due to MDR as the pathogens have become resistant to commercially available drugs, which has triggered the use of more expensive therapies. The rate of success of present-day medical applications like organ transplantation and cancer chemotherapy has contributed immensely towards development of MDR. Differences in the resistance profiles of bacterial and fungal pathogens as well as the quality of public hygiene also have a considerable impact on the effectiveness of antimicrobial agents. Expansion of global trade and tourism lead to increased potential of MDR to spread all over the world and decrease in export and import of various products affecting the economy of developing countries.

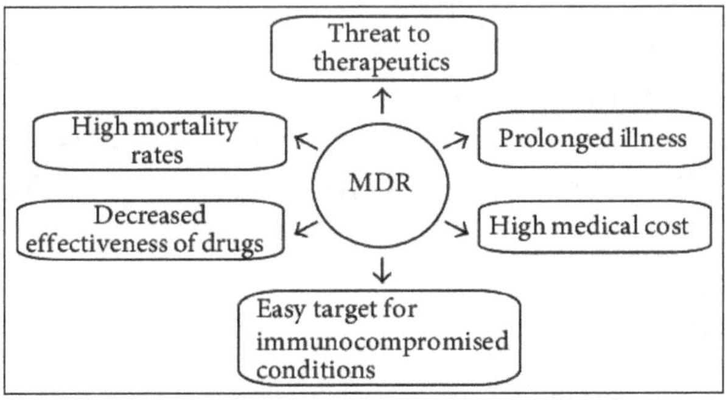

Problems associated with MDR.

Classification of MDR

Despite of administration of appropriate doses of drugs for a specific duration of time, survival of various microbial strains recommends the high levels of resistance developed in them. This clinical failure is due to not only the antimicrobial resistance but also the suppressed immune function, poor/deprived drug bioavailability, or increased rate of drug metabolism. Persistence of microbes after conventional/standard treatments points out different types of antimicrobial drug resistance which is an expanding problem in medical world. MDR can be classified as primary or secondary resistance.

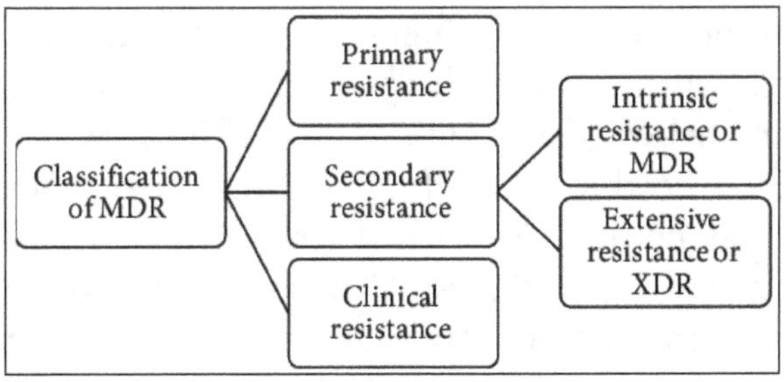

Classification of MDR.

Primary Resistance

It occurs when the organism has never encountered the drug of interest in a particular host.

Secondary Resistance

Also known as "acquired resistance," this term is used to describe the resistance that only arises in an organism after an exposure to the drug. It may further be classified as follows:

- Intrinsic resistance: It refers to the insensitivity of all microorganisms of a single species to certain common first-line drugs, which are used to treat diseases based on the clinical evidence of the patient. It is also known as multidrug resistance (MDR), for example, Mycobacterium tuberculosis to rifampicin and isoniazid or Candida spp. to fluconazole.
- Extensive resistance: It defines the ability of organisms to withstand the inhibitory effects of at least one or two most effective antimicrobial drugs. Also termed as XDR, this seemed to arise in patients after they have undergone a treatment with first line drugs, for example, XDR-TB resistance against fluoroquinolone.

Clinical Resistance

In addition to the above-mentioned types, clinical resistance is defined by the situation in which the infecting organism is inhibited by a concentration of an antimicrobial agent that is associated with a high likelihood of therapeutic failure or reappearance of infections within an organism due to impaired host immune function. In other words, the pathogen is inhibited by an antimicrobial concentration that is higher than could be safely achieved with normal dosing.

Mechanisms of MDR

Resistance is the term referred to as the insensitivity of a microbe to an antimicrobial drug when compared with other isolates of the same species. Although several new drugs have been introduced commercially, this development of resistance among infectious microorganisms is increasing especially in patients under prolonged drug exposure. Antimicrobial drugs generally act on the microbes either by inhibiting a metabolic pathway like nucleotide synthesis which in turn leads to the inhibition of DNA/RNA synthesis and further protein synthesis and disruption of the cell membrane or by competing with the substrate of any enzyme involved in cell wall synthesis (e.g., chitin synthase). Microorganisms have evolved a multitude of mechanisms to overcome the effectiveness of drugs, thereby surviving exposure to the drugs.

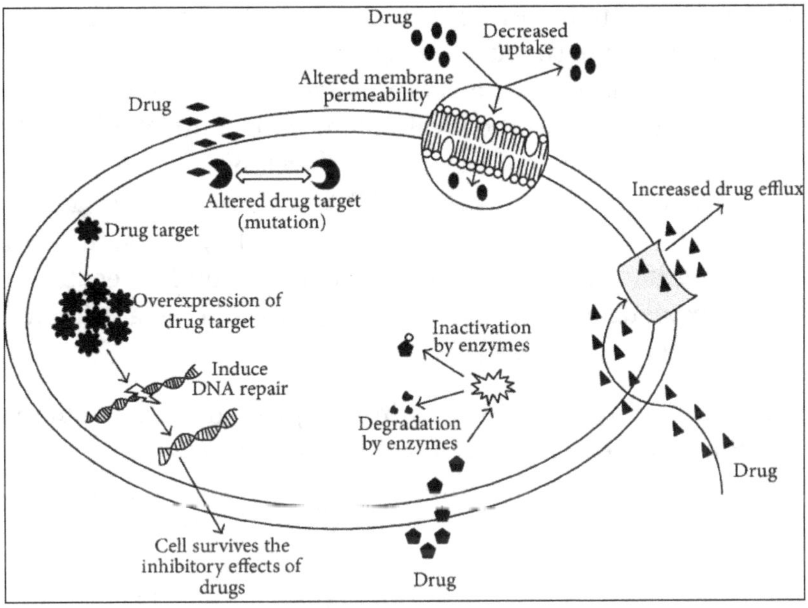

Mechanisms of MDR.

Cell wall, in both bacteria and fungi, plays a crucial role in their survival. As discussed above, drugs inhibit the cell wall synthesis by binding with the peptidoglycan layer in bacteria or affecting ergosterol synthesis (e.g., polyenes) in fungi, thus, blocking the cell growth and division. These organisms undergo certain chromosomal mutations or exchange of extrachromosomal DNA elements through conjugation or transformation (horizontal gene transfer) such as in K. pneumoniae, which can cause alteration in the cell membrane composition (e.g., a reduction in the ergosterol content in fungal plasma membrane) resulting in decreased permeability and uptake of drugs into the cell. Altered membrane composition (such as β-1,3-glucan and lipid content in fungal cell membrane) also leads to lack of active target sites for the drugs (e.g., echinocandins in fungi) to bind. Mutations in the genes encoding for the target cause modifications at the molecular level and retain cellular function by reducing susceptibility to inhibition.

Another mechanism of MDR was found to be an overexpression of drug target enzymes leading to target bypass due to modification in certain metabolic pathways (e.g., azoles and allylamines in fungi), which causes production of alternate target molecules and interference in some protein synthesis. This can influence the access of drugs to the target sites.

Inactivation or enzymatic degradation of antimicrobials by hydrolysis of ester or amide bonds (such as resistance to β-lactams due to β-lactamases, etc.) and chemical transformation of these compounds by acetylation, phosphorylation, adenylation, glycosylation, and hydroxylation have also become increasingly apparent as causes of MDR. The resistant strains of clinical isolates of different microorganisms have developed the ability to oxidize or reduce the antimicrobial compounds to prevent their interaction with the respective targets. Antiviral drugs usually target viral DNA polymerase having the reverse transcriptase activity to inhibit the viral replication. Drug resistant mutant strains undergo mutations in the reverse transcriptase domains of the polymerase gene which affects the interaction between the drug and the enzyme. Resistance to the inhibitory effects of drug on the enzyme can also emerge due to any conformational changes or altered binding of substrate to the viral polymerase. With the lack of effective antiparasitic vaccines yet in sight and new drugs developing slowly, MDR in parasites is emerging as a global public health threat. These parasites such as Plasmodia spp. and Toxoplasma gondii, like bacteria or fungi, also undergo certain point mutations/ substitutions resulting in altered drug targets, alter calcium homeostasis in endoplasmic reticulum, and expel drugs (e.g., chloroquine, atovaquone, antifolate combination drugs, and artemisinin) out of the cells.

MDR mediated by drug efflux pumps remains the predominant mechanism of MDR. The overexpression of genes encoding for ATP-binding cassette (ABC) transporter membrane proteins (e.g., P-glycoprotein (Pgp)), also known as the multidrug efflux pumps which are responsible for the export or expulsion of drugs out of the cell, usually generates MDR and continues cellular functions without any interference. Overexpression of P-glycoprotein, in Entamoeba spp. and Leishmania spp. membrane or multidrug resistant proteins (MRP), affects the fluidity and permeability, leading to an ATP-dependent efflux of the antimicrobials and decreasing their intracellular concentration. MDR is also employed by cancer cells, which limits the long-term use of chemotherapy. An insight into the mechanisms involved in the chemoresistance, which can occur either at the beginning of the therapy (innate) or during the course of treatment, reveals that the cancer cells exhibit overexpression of certain multidrug resistance proteins (e.g., MRP and Pgp) which induce DNA repair mechanism, inhibit apoptosis, alter drug targets, and modify cell membrane composition as well as promoting an increased efflux of drugs preventing proper diffusion into the cells.

Remedies for MDR

The development of MDR is a complicated issue which has become an international dreadful concern. To decrease the rise and spread of MDR, cooperative efforts are requisite because diseases which were curable earlier are becoming major causes of deaths in this era. Moreover, focusing on areas which are susceptible to inappropriate use of antimicrobials by implementation of antibiotic stewardship (defined as coordinated interventions designed to improve and measure the appropriate use of antimicrobials) is the need of the hour. In fact, various antimicrobial stewardship programs (ASPs) are conducted nowadays to optimize antimicrobial therapy, reduce treatment-related cost, improve clinical outcomes and safety, and minimize or stabilize MDR. Interventions through ASPs are either by restricting the availability of selected antimicrobial agents, known as "front-end," or by examining broad spectrum use of antibiotics and then streamlining or discontinuing it, known as "back-end". Therefore, there is an urgent need of support and coordination at the global, regional, subregional, and national level to serve in future progress.

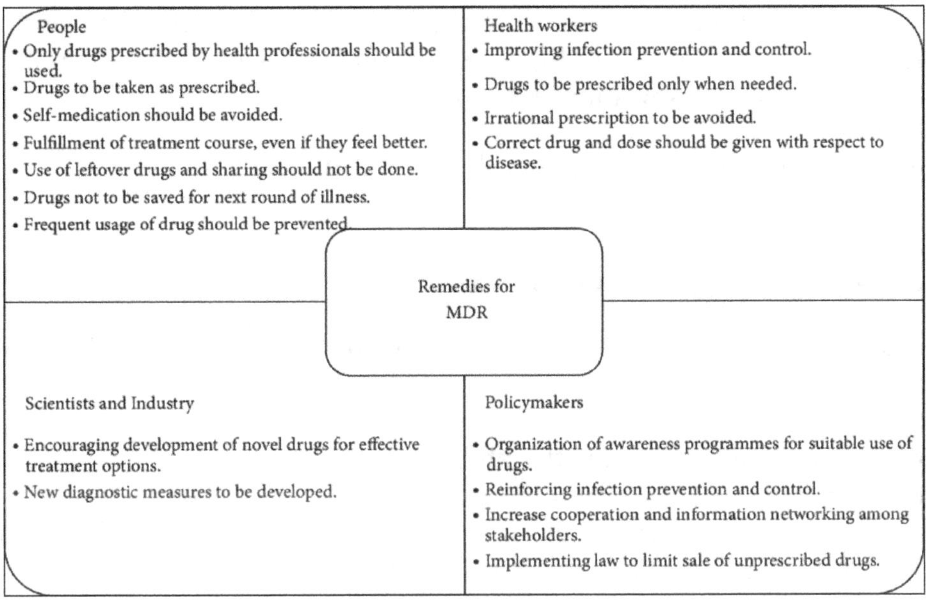

Remedies of MDR.

MULTIDRUG TOLERANCE

Multidrug tolerance or antibiotic tolerance is the ability of a disease-causing microorganism to resist being killed by antibiotics or other antimicrobials. It is mechanistically distinct from multidrug resistance: It is not caused by mutant microbes, but rather by microbial cells that exist in a transient, dormant, non-dividing state. Microorganisms that display multidrug tolerance can be bacteria, fungi or parasites.

Relevance to Chronic Infections

Multidrug tolerance is caused by a small subpopulation of microbial cells termed persisters. Persisters are not mutants, but rather are dormant cells that can survive the antimicrobial treatments that kill the majority of their genetically identical siblings. Persister cells have entered a non- or extremely slow-growing physiological state which makes them insensitive (refractory or tolerant) to the action of antimicrobial drugs. When such persisting microbial cells cannot be eliminated by the immune system, they become a reservoir from which recurrence of infection will develop. Such non-growing bacteria have been observed to persist within mammalian cells. Indeed, it appears that persister cells are the main cause for relapsing and chronic infections. Chronic infections can affect people of any age, health, or immune status.

Medical Importance

Bacterial multidrug or antibiotic tolerance poses medically important challenges. It is largely responsible for the inability to eradicate bacterial infections with antibiotic treatment. Persister cells are highly enriched in biofilms, and it has been suggested that this is the reason that makes biofilm-related diseases so hard to treat. Examples are chronic infections of implanted medical devices such as catheters and artificial joints, urinary tract infections, middle ear infections and fatal lung disease.

Distinction from Multidrug Resistance

Unlike resistance, multidrug tolerance is a transient, non-heritable phenotype. Multidrug tolerant persister cells are not antibiotic resistant mutants. Resistance is caused by newly acquired genetic traits (by mutation or horizontal gene transfer) that are heritable and confer the ability to grow at elevated concentrations of antimicrobial drugs. In contrast, tolerant bacteria have the same MIC (Minimum Inhibitory Concentration) as susceptible bacteria and differ in the duration of the treatment that they can survive. Multidrug tolerance can be caused by a reversible physiological state in a small subpopulation of genetically identical cells, similar to a differentiated cell type. It enables this small subpopulation of microbes to survive the antibiotic killing of their surrounding siblings. Persisting cells resume growth when the antimicrobial agent is removed, and their progeny is sensitive to antimicrobial agents.

Molecular Mechanisms

The molecular mechanisms that underlie persister cell formation and multidrug tolerance are largely unknown. Persister cells are thought to arise spontaneously in a growing microbial population by a stochastic genetic switch, although inducible mechanisms of persister cell formation have been described. Owing to their transient nature and relatively low abundance, it is hard to isolate persister cells in sufficient numbers for experimental characterization, and only a few relevant genes have been identified to date. The best-understood persistence factor is the *E. coli* high persistence gene, commonly abbreviated as *hipA*.

Although tolerance is widely considered a passive state, there is evidence indicating it can be an energy-dependent process. Persister cells in *E. coli* can transport intracellular accumulations antibiotic using an energy requiring efflux pump called TolC.

A persister subpopulation has also been demonstrated in budding yeast *Saccharomyces cerevisiae*. Yeast persisters are triggered in a small subset of unperturbed exponentially growing cells by spontaneously occurring DNA damage, which leads to the activation of a general stress response and protection against a range of harsh drug and stress environments. As a result of the DNA damage, yeast persisters are also enriched for random genetic mutations that occurred prior to the stress, and are unrelated to the stress survival

MULTIDRUG RESISTANCE IN BACTERIA

Large amounts of antibiotics used for human therapy, as well as for farm animals and even for fish in aquaculture, resulted in the selection of pathogenic bacteria resistant to multiple drugs. Multidrug resistance in bacteria may be generated by one of two mechanisms. First, these bacteria may accumulate multiple genes, each coding for resistance to a single drug, within a single cell. This accumulation occurs typically on resistance (R) plasmids. Second, multidrug resistance may also occur by the increased expression of genes that code for multidrug efflux pumps, extruding a wide range of drugs.

The discovery of penicillin in 1928 was followed by the discovery and commercial production of many other antibiotics. We now take for granted that any infectious disease is curable by antibiotic therapy. Antibiotics are manufactured at an estimated scale of about 100,000 tons annually worldwide, and their use had a profound impact on the life of bacteria on earth. More strains of pathogens have become antibiotic resistant, and some have become resistant to many antibiotics and chemotherapeutic agents, the phenomenon of multidrug resistance.

Indeed, some strains have become resistant to practically all of the commonly available agents. A notorious case is the methicillin-resistant Staphylococcus aureus (MRSA), which is resistant not only to methicillin (which was developed to fight against penicillinase-producing S. aureus) but usually also to aminoglycosides, macrolides, tetracycline, chloramphenicol, and lincosamides. Such strains are also resistant to disinfectants, and MRSA can act as a major source of hospital-acquired infections. An old antibiotic, vancomycin, was resurrected for treatment of MRSA infections. However, transferable resistance to vancomycin is now quite common in Enterococcus and found its way finally to MRSA in 2002, although such strains are still rare.

An even more serious threat may be the emergence of gram-negative pathogens that are resistant to essentially all of the available agents. Research had time to react against the threat by MRSA. Thus, there are newly developed agents that are active against vancomycin-resistant MRSA, such as linezolid and quinupristin/dalfopristin. However, the emergence of "pan-resistant" gram-negative strains, notably those belonging to Pseudomonas aeruginosa and Acinetobacter baumanii, occurred more recently, after most major pharmaceutical companies stopped the development of new antibacterial agents. Hence, there are almost no agents that could be used against these strains, in which an outer membrane barrier of low permeability and an array of efficient multidrug efflux pumps are combined with multitudes of specific resistance mechanisms.

Multidrug resistance in bacteria occurs by the accumulation, on resistance (R) plasmids or transposons, of genes, with each coding for resistance to a specific agent, and/or by the action of multidrug efflux pumps, each of which can pump out more than one drug type.

Biochemical Mechanisms of Resistance

Mutational Alteration of the Target Protein

Man-made compounds, such as fluoroquinolones, are unlikely to become inactivated by the enzymatic mechanisms. However, bacteria can still become resistant through mutations that make the target protein less susceptible to the agent. Fluoroquinolone resistance is mainly (but not exclusively) due to mutations in the target enzymes, DNA topoisomerases. Whether resistance of this type is easily transferred to other cells on plasmids depends on the mode of the drug's action. With fluoroquinolones, which will kill the bacterial cell containing any drug-susceptible enzyme, the addition of the gene coding for a drug-resistant enzyme will not make the bacteria completely resistant, and plasmid-mediated transfer of the mutated target gene is unlikely to occur. Nevertheless, the mutants will become more and more prevalent by clonal selection in the presence of selective pressure.

When high-level streptomycin resistance is selected in Escherichia coli in the laboratory, mutation in one of the ribosomal proteins (RpsL) is usually selected. Again, bringing in this mutant gene on a plasmid will not make the recipient resistant because the drug action on the susceptible ribosome (containing the chromosomally coded intact RpsL) will result in cell death. This type of resistance is almost never found in the streptomycin-resistant E. coli of clinical origin, presumably because resistance caused by the drug modification is effectively expressed on plasmids. However, ribosomal resistance mutation is often found in the aminoglycoside-resistant clinical strains of Mycobacterium tuberculosis.

Another example of resistance attributable to target modification is that conferred by the erm gene, which is usually plasmid coded and produces the methylation of adenine at position 2058 of the 50S rRNA, causing resistance to macrolides (erythromycin and many others), lincosamide, and streptogramin of group B, the MLS phenotype. The molecular basis of this phenotype was elucidated through the crystal structure of 50S ribosomal subunit.

Sulfa drugs (synthetic competitors of p-aminobenzoic acids that inhibit dihydropteroate synthetase and trimethoprim, a synthetic inhibitor of dihydrofolate reductase) have been used in combination. They select for drug-resistant mutants of the respective enzymes. In this case, the high-level production of drug-resistant target enzymes from plasmids can make the bacteria resistant, and the resistant genes have spread widely on plasmids.

Enzymatic Inactivation of the Drug

This is a common resistance mechanism for antibiotics of natural origin, such as aminoglycosides (kanamycin, tobramycin, and amikacin), which are inactivated by enzymatic phosphorylation [by aminoglycoside phosphoryltransferase (APH)], acetylation [by aminoglycoside acetyltransferase (AAC)], or adenylation (by aminoglycoside adenyltransferase or nucleotidyltransferase), and β-lactams (penicillins, cephalosporins, and carbapenems such as imipenem), which are inactivated by enzymatic hydrolysis by β-lactamases, usually in the periplasm. Genes coding for these inactivating enzymes can easily produce resistance as additional genetic components on plasmids.

Aminoglycosides

Aminoglycosides are inactivated by modifications that reduce the net positive charges on these polycationic antibiotics. There are now many dozens of aminoglycoside-modifying enzymes known; for example, AAC (3)-II designates an aminoglycoside acetyltransferase acting on position 3 of the substrate and belonging to the second phylogenic grouping among these enzymes. It was proposed in 1973 that a major source of antibiotic-inactivating enzymes is the antibiotic-producing microorganisms, which must protect themselves. Indeed APH (3′)-I, a plasmid-coded aminoglycoside phosphorylase, which is present in 46% of the aminoglycoside-resistant gram-negative bacteria, is strongly related to a chromosomally coded enzyme in the producing strain Streptomyces fradiae. A similar situation also exists for AAC (3)-II, a plasmid-coded enzyme present in more than 60% of aminoglycoside-resistant gram-negative bacteria examined.

After these findings, the pharmaceutical industry developed either semisynthetic or natural aminoglycosides that are not the substrates of common inactivating enzymes. Nevertheless, these drugs are still susceptible to the action of some enzymes, and plasmids found in aminoglycoside-resistant strains in recent years often contain genes coding for several inactivating enzymes.

β-Lactams

Only a few years after the introduction of penicillin into clinical practice, S. aureus developed resistance caused by a β-lactamase coded for it by a plasmid gene. Although this problem was solved by the introduction of methicillin and similar compounds that resist the enzymatic hydrolysis, another enzyme, TEM β-lactamase, was reported in gram-negative bacteria in strains containing multiple-drug-resistant R plasmids that date from 1962. This enzyme became widespread throughout the world, making penicillins with gram-negative activity, such as ampicillin, almost useless. β-Lactamases are classified into several phylogenetic families. Class A includes both the S. aureus and TEM enzymes, whereas Class C represents chromosomally coded enzymes (e.g., AmpC) that are present in many gram-negative bacteria. These two classes are both similar to serine proteases in their mechanism, whereas Class B enzymes are metalloenzymes that hydrolyze carbapenems efficiently.

In response to the spread of β-lactam resistance, various β-lactams were developed. Although the first-generation cephalosporins, such as cephaloridine and cefazolin, were rapidly hydrolyzed by both TEM and AmpC, both cephamycins (such as cefoxitin) and the third-generation cephalosporins containing an oxyimino side chain (such as cefotaxime) were initially reported to resist both types of enzymes. However, the former was inactive against some gram-negative bacteria such as Enterobacter and Serratia. Although the latter was capable of killing these organisms, their introduction into the clinics was followed by the emergence of resistant strains that over-produced the chromosomal AmpC enzyme. In fact, the AmpC enzymes have very low KM values for these compounds, and the values of Vmax/KM were quite high. The AmpC enzyme, however, needs to be induced, and the third-generation cephalosporins were effective against these bacteria simply because they were ineffective inducers of this enzyme. Thus, the third-generation cephalosporins selected for constitutive mutants of ampC. Furthermore, strong expression of plasmid-coded AmpC has been found recently in species that do not express the chromosomally coded ampC.

Subsequently, fourth-generation cephalosporins (cefepime, cefpirome) that are more resistant to hydrolysis by the AmpC enzyme have been developed. However, continued selective pressure resulted in the selection of plasmids that produced mutants of common enzymes, such as TEM or its relative SHV, which can now hydrolyze third- and sometimes even fourth-generation cephalosporins. These enzymes are called ESBL (extended spectrum β-lactamases).

Especially troublesome among the ESBL enzymes are those called CTX-M. The genes coding for these enzymes appear to have originated from the chromosome of an infrequently encountered gram-negative bacterium Kluyvera and have transferred to R plasmids. This transfer or mobilization unusually appears to have occurred many times, and consequently the enzyme rapidly became widespread among R-plasmid-containing pathogenic bacteria.

β-Lactams with a new nucleus, such as carbapenems (e.g., imipenem), still remain quite effective, but their use may eventually result in the increased prevalence of enzymes capable of hydrolyzing these compounds.

Another example of new resistance mechanisms arising through mutations of prevalent resistance genes is a plasmid-coded ciprofloxacin resistance gene. Here an aminoglycoside acetylase, AAC (6′)-Ib, has mutated to acetylate the amino group of ciprofloxacin.

Acquisition of Genes for Less Susceptible Target Proteins from other Species

Sequencing of the genes coding for the targets of penicillin, DD-transpeptidase or penicillin-binding proteins (PBPs), revealed that penicillin resistance among Streptococcus pneumoniae was due to the production of mosaic proteins, parts of which came from other organisms. We note that S. pneumoniae is an organism capable of natural transformation and may import foreign DNA. Interestingly, a similar mechanism of penicillin resistance was also found in another organism capable of natural transformation, Neisseria meningitidis.

An extreme case of this scenario is the generation of MRSA. MRSA strains contain a new methicillin-resistant PBP, called PBP-2A or 2′, whose expression is often induced by methicillin and other β-lactams. The gene for this new PBP is located in a large (30–60-kb) segment of DNA, which apparently came from an organism other than S. aureus and also contains other antibiotic resistance genes. S. aureus is not naturally transformable, and it is unclear how this horizontal transfer of a large DNA segment occurred.

Bypassing of the Target

Vancomycin, a fermentation product from streptomycetes, has an unusual mode of action. Instead of inhibiting an enzyme, it binds to a substrate, the lipid-linked disaccharidepentapeptide, a precursor of cell wall peptidoglycan. Because of this mechanism, many assumed that it would be impossible to generate resistance against vancomycin. However, vancomycin resistance is now prevalent among enterococci, normal inhabitants of our intestinal tract. Because enterococci are naturally resistant to β-lactams, aminoglycosides, macrolides, and tetracycline, these vancomycin-resistant strains of enterococci become prevalent in a hospital environment, colonize the patients, and cause infections that are difficult to treat.

Study of the resistance mechanism showed that the end of the pentapeptide, d-Ala-d-Ala, where vancomycin binds, was replaced in the resistant strain by an ester structure, d-Ala-d-lactic acid, which is not bound by vancomycin. Production of this altered structure requires the participation of several imported genes.

Preventing Drug Access to Targets

Drug access to the target can be reduced locally. It can be also reduced by an active efflux process. In gram-negative bacteria, the access can be reduced generally by decreasing the influx across the outer membrane barrier.

Local Inhibition of Drug Access

Tet(M) or Tet(S) proteins, produced by plasmid-coded genes in gram-positive bacteria, bind to ribosomes with high affinity and apparently change the ribosomal conformation, thereby preventing the association of tetracyclines to ribosomes. Plasmid-coded Qnr proteins, which have become more prevalent in recent years, are thought to protect DNA topoisomerases from (fluoro)quinolones.

Drug-specific Efflux Pumps

Drug resistance owing to active efflux was discovered with the common tetracycline resistance protein TetA in gram-negative bacteria, which catalyze a proton-motive-force-dependent outward pumping of a tetracycine-Mg complex.

Nonspecific Inhibition of Drug Access

In the laboratory where nutrient-rich medium is used, β-lactams often select for porin-deficient mutants. However, the generally decreased outer membrane permeability is somewhat detrimental to the bacterial growth because the nutrient influx is also reduced, and such mutants are not common among clinical specimens. Nevertheless, porin mutants are found in some species of Enterobacteriaceae (Enterobacter aerogenes, Klebsiella pneumoniae) as a means of last resort resistance to the more recent versions of β-lactams that withstand inactivation by common β-lactamases. Mutations within the coding sequences of the porin also have been reported, which possibly reduce the permeation rates of bulky β-lactams without affecting those of smaller nutrient molecules.

Sources of the Resistance Genes

Producing Organisms

some of the aminoglycoside-resistant genes appear to be derived from streptomycetes producing these antibiotics. The genes coding for vancomycin resistance appear to originate in a similar way. Resistance here requires the production of several new enzymes, and it is unlikely that the genes coding for these enzymes evolved in the few decades after vancomycin was introduced. Indeed, the genes in the vancomycin-resistant clinical isolates of enterococci were found to be homologs of those found in the vancomycin-producing streptomycetes, organized in exactly the same manner, an observation that leaves no doubt on the origin of these resistance genes.

Microorganisms in the Environment

Some resistance genes are found in the chromosome of environmental bacteria. A classical case is the ampC gene in the environmental genera of Enterobacteriaceae, such as Enterobacter, Serratia, and Proteus, and in the soil organism P. aeruginosa. These genes do not show signs that they have been imported in the recent past, and it is revealing that the ampC gene of the exclusive animal symbiont E. coli lacks the induction mechanism and that the pathogen Salmonella spp. lacks the ampC gene entirely. In this connection, examination of a random collection of soil-dwelling strains of Streptomyces and their relatives showed that 60% to 100% of them were resistant to several antibiotics tested, suggesting that antibiotic-resistant genes are abundantly present in this habitat. Although most antibiotics may be present in soil only at very low concentrations, the recent discovery of microorganisms that utilize antibiotics as nutrients is suggestive of the evolutionary origin of some antibiotic degradation (resistance) genes.

Assembly, Maintenance and Transfer of Resistance Genes

R plasmids often contain many resistance genes; they are maintained stably in the host strains of bacteria and are transferred very efficiently to neighboring drug-susceptible cells.

Assembly of Resistance Genes in R Plasmids

We have seen that most drug resistance genes are effective when expressed from plasmids. Remarkably, many such genes are often present on a single R plasmid, so that multidrug resistance can be transferred to a susceptible bacterium in a single conjugation event. When the R plasmids were discovered in Japan in the 1950s, many of them already contained resistance genes for aminoglycosides, tetracycline, chloramphenicol, and sulfonamides. Sequencing of many plasmids has now shown how this clustering occurred.

In the sequence of the early-generation R plasmids, we see that most of the resistance genes are components of transposons, which can deliver the genes to any piece of DNA. This is seen in the plasmid R100.

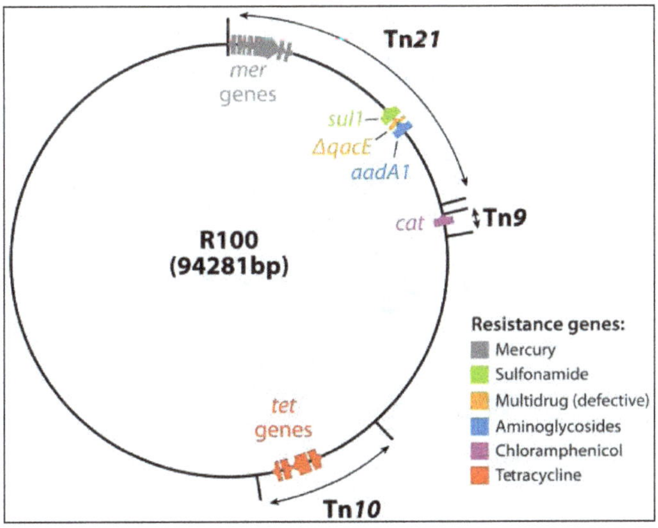

Map of an early R plasmid R100. Tetracycline resistance gene tetA is in the transposon Tn10, chloramphenicol acetyltransferase (cat) as a part of Tn9, and sulfonamide resistance gene sul1 and an aminoglycoside adenyltransferase gene aadA1 as a part of the large transposon Tn21. The resistance genes are color coded, and the direction of transcription is shown by the arrows. Much of the blank areas outside the transposons are occupied by the transfer genes, needed for the formation of relaxosomes, the coupling factor, and the mating-pair formation. This figure is based on the nucleotide sequence deposited by and GenBank sequence NC 002134, submitted by G. Sempei and K. Mizobuchi. ΔqacE is a defective version of a gene coding for a multidrug efflux pump of Smr family.

Tn21 is a particularly remarkable example of large, complex, multiply composite transposons. Interestingly, it contains mercury resistance genes. The sulfonamide resistance and aminoglycoside resistance genes in Tn21 were assembled.

The discovery that many resistance genes in R plasmids contain a unique 59-base

3′-sequence tag led to the discovery of a remarkable apparatus called an integron. An integron contains a gene coding for an integrase, which catalyzes the insertion of resistance genes at a predetermined site downstream from a strong promoter. Once integrated, the resistance gene becomes marked by the tag, so that it can easily become integrated into another integron, perhaps containing a different set of resistance genes. In addition to this advantage of high mobility, the resistance genes when inserted into an integron become organized into a single operon, with the same orientation of transcription under a strong promoter supplied by the integron structure. In the example seen in Tn21, the integron already containing a sulfonamide resistance gene sul1 and a truncated version of a multidrug efflux gene qacE has integrated an aminoglycoside resistance gene aadA1 at the specific integration site attI. An integron may contain up to eight resistance genes.

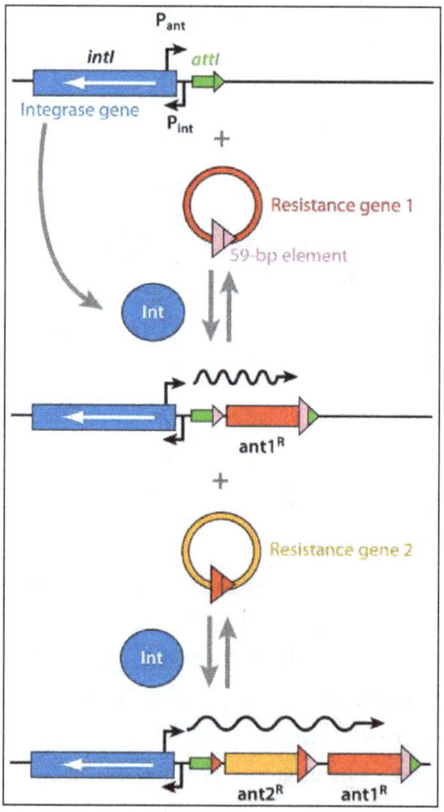

Presumed mechanism of resistance gene capture by integrons. An integron contains the gene for site-specific integrase (blue) and the specific integration site attI (green). When the resistance gene 1 (red) in the circular cassette form containing the 3′-terminal 59-bp element (pink) is available, the gene is integrated at the attI site, regenerating a slightly altered attI sequence (now in green and pink). This can then accept the second resistance gene casette, and the process can go on in this manner. The resistance genes, all in the same orientation, are transcribed by the powerful promoter provided by the integron (P_{ant}).

In addition, many of these integrons carry enzymatic machinery for transposing the entire integron structure to other places. Furthermore, the integron is usually inserted into a larger transposon such as Tn21, which allows the entire array of the resistance genes to hop between different plasmids and between plasmid and chromosome. It is unclear how the modern integrons evolved, but there is a rather similar mechanism that builds the assembly of many genes in Vibrio cholerae.

Recently, many integron structures were found to be associated with a downstream structure called ISCR element, containing a putative transposase gene. It apparently functions in an unusual, open-ended transposition event and recruits various resistance genes and delivers them close to the integron structure, resulting in the assembly of yet more resistance genes.

Maintenance of R Plasmids in the Host Cells

Recombinant plasmids derived from cloning vectors are frequently lost from the host cells even when they exist in relatively high copy numbers. In contrast, most natural R plasmids are remarkably stable and are rarely lost during the multiplication of host cells, even when the copy number is low. This is because R plasmids usually contain genes that ensure the correct partitioning of copies to daughter cells. In addition, some natural plasmids contain the "killer" elements composed of a stable killer protein or mRNA and an unstable inhibitor protein or antisense RNA, so that the loss of the plasmids will result in the death of the host cell.

Because of the presence of these regulatory networks, plasmids sharing a similar replication/partition machinery cannot stably coexist in the same host cell, and this phenomenon has been utilized for the classification of plasmids into "incompatibility groups." In plasmids of enteric bacteria, this classification includes about two dozen such groups.

Cell-to-cell Transfer of R Plasmids

R plasmids are not only stably maintained, but also usually transferred between bacterial cells at a very high efficiency, in many cases approaching 100%. The molecular mechanism of this cell-to-cell transfer has been studied mainly in E. coli and its relatives and in the plasmids of incompatibility group FI (F or sex plasmid), P (RP4), and W. In recent years, the plasmid transfer genes were found to be homologs of the virB genes of Agrobacterium tumefaciens, which transfer a piece of bacterial DNA into a plant nucleus, and several effector protein injection systems of human and animal pathogens. These systems are now called Type IV Secretion Systems. A specific sequence in the circular, double-stranded DNA of plasmids is recognized, and a single-stranded cut is made by the VirD2/TraI enzyme, with the enzyme protein remaining covalently attached to the 5′-end of the cut strand. This complex is sometimes called a relaxosome, as the nicking of one strand results in the relaxation of supercoiling of the plasmid

DNA. The protein-(single-stranded) DNA complex is then exported by the "mating-pair formation" complex, composed of about a dozen different proteins. The export of relaxosomes, rather than protein-free DNA, explains the observation that some Type IV Secretion Systems export pure proteins. Before this happens, however, the leading 5-- end of the exported DNA, with the VirD2/TraI protein attached, must be taken up by an ATPase (VirD4 or TraG). The latter is called a coupling protein because of its role in coupling the export substrate to the export apparatus. Although one of the original strands of the R plasmid DNA is transferred completely into the recipient cells, in the donor cell this strand is replaced by a rolling-circle replication mechanism, so that the complete R plasmids now exist both in donor and recipient cells. One would suspect that the entry of the DNA of foreign origin would be recognized by the strain-specific restriction endonuclease of the recipient cells. However, the initial DNA piece that enters is single stranded (although the complementary strand will be synthesized soon) and thus may escape this mechanism. In addition, some R plasmids are equipped with functions that antagonize the attack by restriction endonucleases.

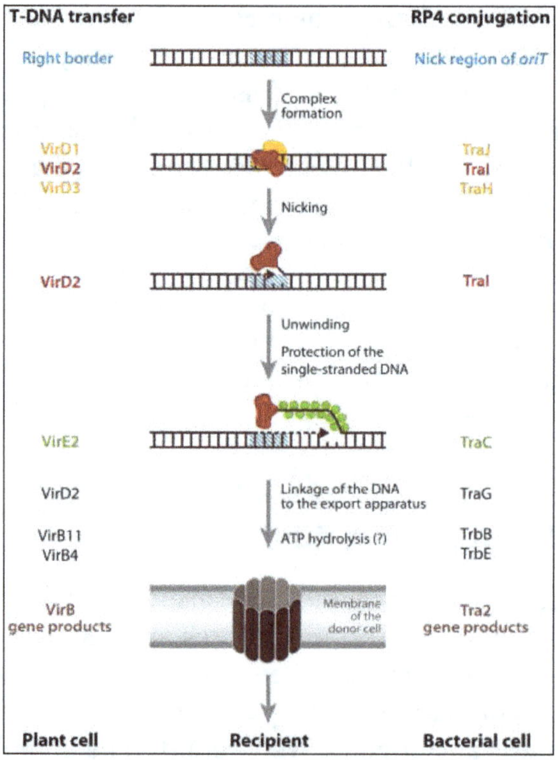

Steps in T-DNA transfer in Agrobacterium tumefaciens (annotation on the left) and conjugative plasmid DNA transfer in enteric bacteria (annotation on the right).

Some R plasmids are very small and do not contain the genes needed for the construction of a mating-pair formation complex. However, these plasmids can be mobilized by the transfer genes of other plasmids and can be transferred efficiently into recipient cells. Finally, it should be emphasized that conjugational transfer of plasmids is not a novel mechanism invented for multidrug resistance; indeed, horizontal transfer of genes, to a large extent through the conjugational mechanism, has been the main mechanism used for the evolution of many groups of bacteria.

Gram-positive bacteria also carry out cell-to-cell transfer of plasmids. In addition, a large transposon of a special class, called conjugative transposons, play a prominent role in these bacteria. They use a λ-phage-like mechanism of transposition, which generates a circular DNA as an intermediate. This intermediate can insert at a predetermined location in the plasmid or chromosomal DNA, but it can also behave like a conjugative plasmid and can transfer its copy into another bacterium by conjugation. Remarkably, in the transposons containing the tetracycline resistance gene, the transfer function is induced strongly by the presence of tetracycline. Thus, in these transposons, the drug resistance genes are not just passive passengers but are a part of an integrated regulatory mechanism of conjugational transfer. Such transfer also occurs easily between distantly related organisms, for example between gram-positive and gram-negative bacteria.

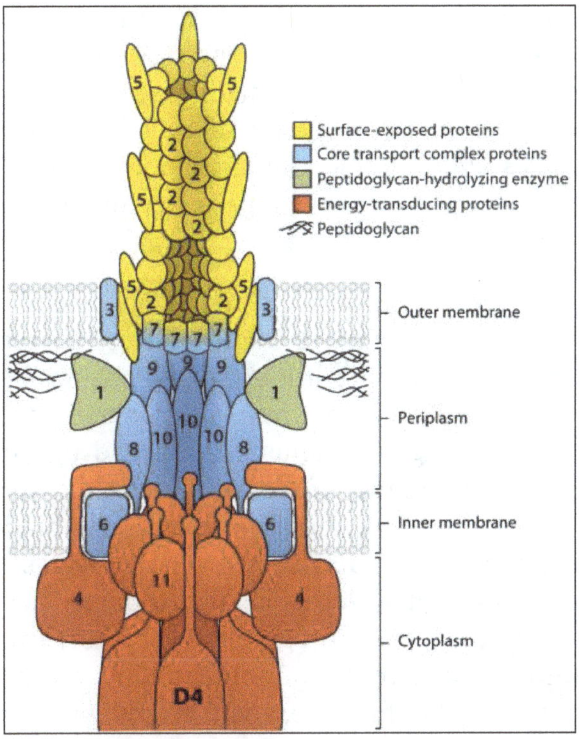

A model of the mating-pair formation complex with the coupling protein (VirD4). The proteins are named after the VirB components in Agrobacterium tumefaciens, but homologs for most of these proteins are found in the conjugative R plasmid transfer systems. Number n denotes the product of virBn gene. The virD4 product is shown as D4.

Multidrug Efflux Pumps

The active efflux of drugs has been known to play a predominant role in the resistance to certain individual drugs, such as tetracycline, Analysis of S. aureus strains that were resistant to multiple cationic bacteriocides and were causing hospital-acquired infections showed, however, that these strains contained plasmids coding for a multidrug efflux transporter QacA (or QacB), belonging to the Major Facilitator superfamily (MFS), the first multidrug efflux pump identified in bacteria. Since then, multidrug efflux pumps belonging to various families have been discovered, and the contribution of such pumps to multidrug resistance became clear. Structures and possible mechanisms of these transporters have been reviewed recently

Multidrug efflux pumps belonging to the major facilitator superfamily:

MFS Pumps with 14 Transmembrane Segments

QacA and QacB were the first examples of this class. These pumps actively extrude monocationic biocides and dyes, such as benzalkonium chloride, cetyltrimethylammonium bromide, and ethidium bromide, and in addition, QacA pumps out dicationic biocides, such as chlorhexidine and pentamidine isethionate. Each protein contains transmembrane segments (TMSs) with several acidic amino acid residues in the transmembrane region. Comparison of these two protein sequences and site-directed mutagenesis showed that the extrusion of dicationic compounds by QacA, but not by QacB, depended upon the presence of an aspartate residue in the TMS 10.

The expression of QacA pump is negatively regulated by the QacR repressor, and binding of some of the QacA substrates to QacR relieves this repression, resulting in the increased expression of the pump. Although QacA has not been crystallized, crystallographic studies of the QacR-inducer complex gave us insights on how a binding site can accommodate diverse ligands. Thus, the QacR drug-binding pocket is large (1100Å3), and many drugs bind to one of the "minipockets" within this cavity. Unlike the binding of most ligands to conventional enzymes, the binding involves few hydrogen bonds and relies mostly on stacking and van der Waals interactions, with the charges of the ligands usually neutralized by acidic residues. These features are similar to the binding of ligands to the human pregnane X receptor, a transcriptional regulator that binds to a number of xenobiotics. QacR can bind two ligands simultaneously to different minipockets of the binding site, and a given ligand can be bound in different ways within the binding cavity. Another regulatory protein, TtgR, belonging to the same family as QacR but regulating a transporter in the Resistance-Nodulation-Division (RND) family, was crystallized with several antibiotics and plant secondary metabolites. The binding pocket here is even larger (1500Å3), and some ligands utilize the broad hydrophobic area of the pocket, whereas phloretin in its high-affinity mode binds to the deep valley within the pocket, utilizing several H-bonding interactions, consistent with the extremely wide range of ligands that regulate the expression of the cognate RND pump.

Another multidrug transporter belonging to this branch is EmrB of E. coli, which confers resistance to uncouplers such as carbonyl cyanide m-chlorophenylhydrazone and to antibiotics such as nalidixic acid and thiolactomycin. EmrB is coded for by a chromosomal gene, together with EmrA, a periplasmic adaptor protein that connects the pump to an outer membrane channel, TolC, so that the drugs can be exported directly into the medium.

MFS Pumps with 12 Tmss

The chromosomally coded NorA of S. aureus, in contrast, belongs to a branch with 12 TMS transporters. NorA produces resistance to fluoroquinolones, as well as to cationic dyes and cationic inhibitors including puromycin and tetraphenylphosphonium. The S. aureus chromosome contains at least two other homologs of NorA, NorB and NorC, which produce a similar phenotype. Because all these efflux pumps are inhibited by reserpine, sensitization of bacteria to substrate drugs in the presence of reserpine can be used as an effective tool to assess the contribution of the efflux process to resistance. Thus, in about one-half of S. aureus strains isolated from the bloodstream of patients, efflux made a strong contribution to resistance, and more than half of these strains overexpressed the three chromosomally coded pumps just mentioned.

Some MFS pumps have been studied as model systems, although they have little relevance in the clinic. LmrP of Lactococcus lactis pumps out cationic dyes, daunomycin, tetracyclines, and macrolides, and MdfA of E. coli can confer increased resistance to cationic dyes, chloramphenicol, and fluoroquinolones when overexpressed from plasmids in a mutant strain lacking the constitutive RND pump AcrB, although mdfA deletion produces no effect on drug susceptibility. Study of the LmrP pump showed that fluorescent dyes are captured from the inner leaflet of the bilayer, the pump thus acting as a "vacuum cleaner" of the membrane. The MdfA pump is remarkably versatile, for example, pumping out neutral substrates in an electrogenic manner while treating cationic substrates in an electroneutral way. Another unexpected finding was that the mdfA deletion mutants were hypersensitive to alkaline pH, presumably because MdfA functions as a K+/H+ antiporter. This is reminiscent of the similar function for a chromosomally coded "monodrug efflux pump," TetB(L), of Bacillus subtilis.

Among MFS multidrug efflux pumps, the crystal structure was solved for EmrD of E. coli, which pumps out uncouplers, quaternary ammonium biocides, as well as sodium dodecylsulfate. The central cavity, a likely substrate-binding site, is lined with side chains of both an aliphatic and an aromatic nature.

Multidrug Efflux Pumps of the small Multidrug Resistance Family

Efflux pumps of a very different structure, belonging to the Small Multidrug Resistance (SMR) family, were first discovered coded on S. aureus plasmids and were subsequently found to be coded also on the chromosomes of gram-negative bacteria. These proteins, which extrude cationic compounds such as quaternary ammonium biocides or ethidium, represent the smallest transport proteins known, containing only four TMS in a 110-residue sequence in the EmrE of E. coli. EmrE has been studied intensively. There is only one charged residue in the transmembrane segments, Glu14, which has an unusually high pKa of about 8.5. When the transporter encounters the substrate, this residue becomes deprotonated and instead binds the substrate, as shown with the transporter solution in detergents. Thus, the inward flux of proton(s) and the outward flux of the substrate appear to be coupled by the sharing of a common binding site. In terms of the three-dimensional structure, EmrE seems to exist primarily as a symmetrical dimer, according to the biochemical and the electron cryomicroscopy data. Although much confusion ensued following the publications suggesting that the dimer is asymmetric, with one protomer facing the outside and the other facing inside, these papers have now been retracted as a result of calculations using a faulty software.

Multidrug Efflux Pumps of Resistance-nodulation-division Family

Resistance-nodulation-division pump usually exists as a part of a tripartite complex: Efflux pumps of this superfamily (such as AcrB of E. coli and MexB of P. aeruginosa) play an important role in producing multidrug resistance in gram-negative bacteria. This is because these pumps become associated with two other classes of proteins, the outer membrane channel exemplified by TolC of E. coli, belonging to the OMF (outer membrane factor) family of proteins and the periplasmic "adapter" protein such as AcrA of E. coli and MexA of P. aeruginosa, classified into the MFP (membrane fusion protein) family. The association of these proteins was confirmed by chemical cross-linking and by the isolation of the complex containing all three proteins. As was emphasized already in 1996, this construction allows the direct export of drugs into the external medium, rather than into the periplasmic space. This is a huge advantage because once exported into the external space drug molecules must traverse the outer membrane barrier to reenter bacterial cells. Thus, these pumps work synergistically with the outer membrane barrier. Wild-type strains of most gram-negative bacteria are resistant to most lipophilic antibiotics (for E. coli, they include penicillin G, oxacillin, cloxacillin, nafcillin, macrolides, novobiocin, linezolid, and fusidic acid), and this "intrinsic resistance" was often thought to be caused by the exclusion of drugs by the outer membrane barrier. Indeed, breaching the outer membrane barrier does sensitize E. coli cells to the drugs just mentioned. However, the inactivation of the major RND pump AcrB of E. coli makes the bacteria almost completely susceptible to these agents [the minimal inhibitory concentration (MIC) of a lipophilic penicillin, cloxacillin, goes down from 512 μg/ml in the wild type to only 2 μg/ml] even in the presence of the intact outer membrane barrier. Thus, the characteristic intrinsic resistance of gram-negative bacteria owes as much to the RND pumps as to the outer membrane barrier.

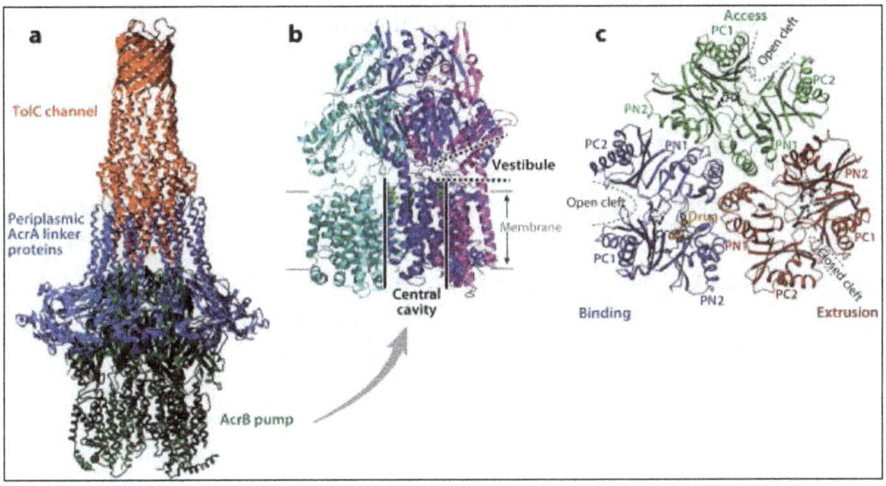

(a) The model of the AcrB-AcrA-TolC tripartite complex that exports drugs directly into the medium. The transmembrane domain of the AcrB pump trimer is embedded in the cytoplasmic membrane, whereas its periplasmic domain is connected to the TolC channel through a number of periplasmic AcrA linker proteins. From Reference with permission from Elsevier Ltd. (b) AcrB trimer. Each protomer is shown in a different color. The large central cavity (thick black lines) is connected to the periplasm through vestibules (thick dotted lines) between protomers. The proximal portion of the structure was cut away to reveal the presence of vestibule. Drawn by using PyMol with Protein Data Bank coordinate file 1OYE. (c) The periplasmic domain of the asymmetric AcrB trimer viewed from the top. The conformation of each protomer is characteristic, with open or closed external clefts. A drug molecule is seen bound to the binding protomer. From Reference with permission from Nature Publishing Group.

Some RND pumps show extremely wide substrate specificity. E. coli AcrB can pump out not only most of the common antibiotics but also dyes, detergents, and even solvents. Although AcrB cannot pump out aminoglycosides, there are homologs such as AcrD that carry out this function. Most gram-negative bacteria contain several chromosomal genes coding for such pumps, and their expression may become increased through regulatory responses or mutations, making the bacteria more resistant to practically all antimicrobial agents in one single step.

Finally, at least some RND pumps appear to capture the drugs either in the periplasmic space or from a location that is in equilibrium with the periplasm. This was predicted because carbenicillin, which cannot cross the inner (cytoplasmic) membrane owing to the presence of two carboxylate groups, is still a good substrate for these pumps. This allows the pumps not only to prevent the entry of drugs into the cytoplasm, but also to extrude agents, such as β-lactams, that have targets in the periplasm. Consistent with this concept, the substrate specificity of RND pumps is primarily determined by their large periplasmic, extramembrane domains, as shown by study of chimeric transporters or mutants with altered substrate specificity. Importantly, the periplasmic capture mode allows the RND tripartite pumps to collaborate synergistically with simple efflux pumps that extrude substrate drugs only into the periplasm.

Regulation of RND Pump Expression

Expression of many RND pumps is repressed by specific repressor proteins. Sometimes, the pump expression is increased by substrates, for example, MexXY, an aminoglycoside efflux pump of P. aeruginosa, is induced by its substrates. However, the mechanism of induction is more complex than with that of QacA, where the repressor QacR is inactivated by the binding of the inducer drug. Thus, induction requires the inhibition of ribosomal function, which is not necessarily caused by aminoglycosides, and apparently involves another protein.

With most other systems, the pump gene expression is not regulated by the substrates, and it is not known what ligands, if any, become bound to the local repressors. Thus, the efflux pump overproduction in clinical isolates occurs frequently as a result of mutation in the repressor gene, as with mexR mutations leading to overproduction of MexAB-OprM system in P. aeruginosa. A fraction of such MexAB-overproducing strains has no changes in the mexR gene and has mutations in the gene PA3721, the so-called nalC phenotype. Finally, the nalD gene, located upstream from the mexAB-oprM operon, also is involved in regulation. The mexCD-oprJ operon is again under the control of the NfxB repressor, coded by an upstream gene, and the pump overproducers have mutations in the repressor gene. The mexEF-oprN operon is exceptional in being regulated by an upstream activator, MexT. MexEF-OprN-overproducing strains do not contain mutations in the mexT gene, and instead have mutations in PA2491, a gene coding for an enzyme-like molecule.

The expression of acrAB operon in E. coli is repressed by its cognate repressor AcrR, but the small effector molecules that bind to this repressor are not known. Instead, the transcription of the acrAB operon is mainly controlled by the AraC family global activators MarA, SoxS, and Rob. MarA and SoxS are exceptionally small proteins containing only the DNA-binding domain, and thus, regulation occurs at the level of production of these proteins. The MarA production is regulated by a repressor MarR, which becomes inactivated by salicylate and plumbagin. Interestingly, transketolase A, whose production is elevated under oxidative stress, binds to MarR and relieves its repression of MarA production. The level of active SoxS repressor in the cell is determined by its own repressor SoxR, which contains Fe-S centers and becomes inactivated by superoxide. Rob is constitutively expressed and is also much larger than MarA and SoxS, containing a domain outside the DNA-binding domain. Indeed, Rob becomes inactivated by binding substrate-like molecules, such as α,α′-dipyridyl, fatty acids, and bile salts.

Expression of some RND-type transporters, such as MdtBC, is regulated by a two-component system, BaeSR. Indole upregulates MdtBC as well as AcrD through BaeSR, although it is unclear if indole is directly sensed by the sensor kinase BaeS.

Plasmid-driven overproduction of SdiA, the soluble receptor for the quorum-sensing signal acyl-homoserine lactone, increased expression of AcrAB. However, growth of E. coli in the presence of hexanoyl-homoserine lactone did not upregulate the acrAB operon.

Biochemical and Crystallographic Studies

Biochemical studies of the RND-type efflux pumps were hampered by the fact that the pumps function as multiprotein complexes spanning two membranes. Nevertheless, the basic kinetic constants of the AcrB pump were recently determined. When β-lactams are added to intact E. coli cells, they diffuse across the outer membrane and, in periplasm, are either hydrolyzed by the periplasmic β-lactamase or extruded by the AcrAB-TolC pump. Because the kinetic constants of the β-lactamase are known, we can calculate the periplasmic drug concentration from the hydrolysis rate. The efflux rate at that periplasmic concentration is obtained as the difference between the rate of influx (following Fick's first law of diffusion) and the hydrolysis rate. Data with nitrocefin suggests saturation kinetics of efflux with a half-saturation concentration of ~5 μM and a turnover number about 10. Although nitrocefin did not show positive cooperativity, such a behavior was evident with cephalothin, cephamandole, and cephaloridine, which showed much higher K0.5 values than nitrocefin did. The positive cooperativity is consistent with the functionally rotating mechanism of AcrB function.

Another biochemical approach used was the reconstitution of purified RND transporter into proteoliposomes. Because most of the substrates of these pumps are lipophilic and spontaneously cross the bilayer, the assay used the movement of labeled phospholipid from the exporter-containing vesicle to acceptor vesicles that were devoid of proteins. the AcrB pump functions as a proton antiporter and also showed that, among the substrates, conjugated bile salts had the highest affinity to AcrB. Another result was that the addition of AcrA increased very strongly the rate of transport. Although this was interpreted as the result of connection between vesicles mediated by AcrA, AcrA is essential in producing the aminoglycoside pumping activity of reconstituted AcrD where such juxtaposition of vesicles is not needed, and AcrA probably activates directly the pumping activity of AcrB.

Most important pieces of knowledge concerning the RND pump complex came from crystallography. Thus, the structure of the outer membrane channel TolC was solved in 2000, followed by the structure of its P. aeruginosa homolog, OprM. These proteins exist as a tightly woven trimer, containing a single 12-stranded β-barrel traversing the outer membrane and a remarkably long (~70-Å) periplasmic extension of the channel in the form of long α-helical bundles. Murakami et al. solved the crystallographic structure of AcrB trimer in 2002, which was a first for a proton antiporter. The periplasmic portion is at least as large as the transmembrane portion, and the top of the periplasmic domain, the TolC-binding domain, has a dimension that is similar to the tip of the α-helical bundle of TolC. This was followed by the crystallographic elucidation of the structures of the central portion (about two-thirds) of the adaptor proteins MexA and AcrA. These are elongated proteins, as shown in 1999, and in crystals show a strong tendency to become packed side by side to form hexamers and heptamers. Thus, the structures of all three component classes of the tripartite assembly are known, and it is possible to propose how these proteins are assembled together. In such models, the coiled-coil domain of the adaptor is assumed to interact with the outer membrane channel protein, a hypothesis that is supported by biochemical data. Studies with chimeric constructs of AcrA homologs showed that the C-terminal domain, whose structure is not known from crystallography, is essential for interaction with AcrB. Mutant studies also suggest that the domain close to this end of AcrA, containing the $\alpha + \beta$ structure, is the main site of its interaction with the periplasmic domain of RND pumps.

The outer membrane component TolC and its homologs are first synthesized with the signal sequence and then are exported into the outer membrane. The periplasmic adaptors are either lipoproteins with the covalently bound lipid at its N terminus or contain a hydrophobic N-terminal helix and are thought to become bound to the outer surface of the inner membrane. The RND pump itself, as a classic example of polytopic inner membrane protein, requires the signal-recognition particle and the SecY apparatus for insertion into the membrane. The interaction between the component proteins of the AcrB-AcrA-TolC complex was studied in vitro by titration calorimetry.

The AcrB trimer structure showed that there was a small opening (vestibule) between subunits at the bottom of the periplasmic domain, close to the external surface of the membrane bilayer, and this led to the top of the large central cavity in the transmembrane domain. It was thus hypothesized that the drugs diffused through the vestibule into the central cavity, where they were captured. Indeed, cocrystallization of AcrB with various drugs produced crystals with drug molecules within this central cavity. However, there was no proof that the bound drugs were on their correct path to extrusion, and the extrusion pathways were not clear in the crystal structure. We have attempted to solve this problem by crystallizing AcrB mutants in which one of the residues putatively involved in proton translocation was altered; we assumed that such proteins would assume a conformation mimicking that of a transient intermediate during the process of drug extrusion. There was an extensive conformational alteration in the transmembrane domain, but only minimal alterations were seen in the periplasmic domain, which is likely the main site of substrate binding and extrusion. The reason became clear from the papers describing the asymmetric trimer structure of AcrB, especially as the Murakami group succeeded in solving the structure in which one of the protomers contained a substrate (minocycline or doxorubicin). Each protomer takes a conformation that is unique. Indeed the transmembrane domain of our mutant AcrB was similar in conformation to that of one of the protomers, called the extrusion protomer. However, extensive conformational changes in the periplasmic domain require complementary accommodation by the neighboring protomers. This was not possible in our construct, in which all protomers corresponded to the mutant protein. In the binding protomer, the periplasmic domain encloses an expanded binding pocket containing several aromatic residues, and there is an open pathway between the binding pocket and the vestibule. In the extrusion protomer, the binding pocket becomes much narrower, and the pathway to the vestibule becomes closed. These crystallographic observations suggest that the protomers within the trimeric AcrB go through a cyclic conformational change, from the open conformation through the ligand-bound one to finally the extrusion conformer, whose transmembrane domain shows signs of a disrupted network among proton-translocating residues. This cyclic change involves the opening and closing of the large external cleft in the periplasmic domain, and the forced closing of this cleft by fast-acting disulfide cross-linking agents stops the function of the pump instantaneously, thus providing biochemical support for this hypothesis. A similar result was obtained with a somewhat different approach. More recently, we succeeded in producing a giant gene containing three copies of the acrB genes connected together by linker sequences and, by using this method, showed conclusively that inactivation of just one protomer in the trimeric complex completely abolishes the pumping activity, a result supporting the functionally rotating mechanism mentioned above.

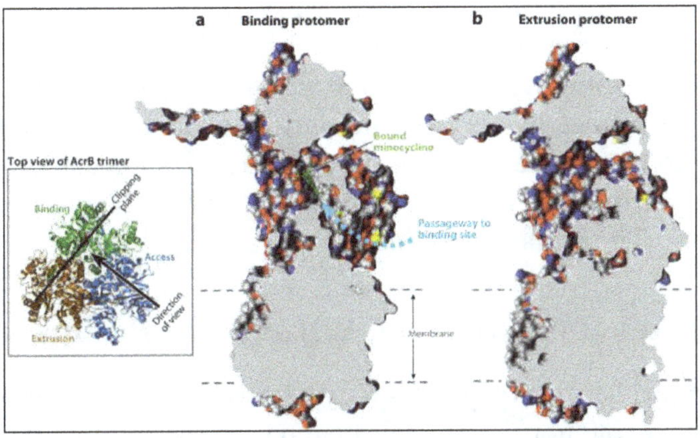

Cut out view of the binding protomer with the bound minocycline (in a ball-and-stick representation in green) (a) and the extruding protomer (b), both from Protein Data Bank file 2DRD. The wide passageway from the external surface to the binding site (dashed arrow) seen in (a) appears to be completely closed in (b). The dashed lines show the approximate limits of the membrane bilayer. The inset shows the location and direction of the clipping plane as well as the direction of the view in (a). The binding, extrusion, and access protomers are indicated.

The binding pocket identified in the asymmetric structure is large and flexible. In fact, the two substrates, which were cocrystalized with AcrB, minocycline, and doxorubicin, predominantly occupy somewhat different parts of the cavity; this explains the extremely wide substrate specificity of this pump. The pocket in AcrB, which favors lipophilic substrates, has a remarkably hydrophobic surface. When AcrD, which pumps out basic and hydrophilic aminoglycosides, is homology modeled following the binding protomer of AcrB, its hypothetical binding pocket is seen to contain many electronegative atoms.

This observation increases our confidence that this is the genuine substrate binding site in these transporters.

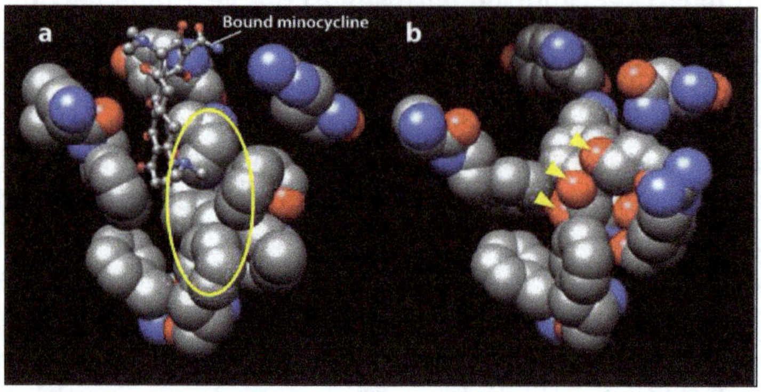

Substrate binding pockets of AcrB and a homology-modeled AcrD. (a) Residues facing the bound minocycline (in a ball-and-stick model), i.e., residues 177, 178, 275–278, 610, 612, 615, 620, 626, and 628 of the binding protomer of AcrB in Protein Data Bank file 2DRD are shown in space-filling models. The walls are remarkably hydrophobic, as seen for example in the area covered by the yellow ellipse. (b) Hypothetical structure of the corresponding region in AcrD. AcrD structure was built by homology modeling, using the binding protomer of AcrB as the template. The area covered by the ellipse in (a) is now seen to be studded with oxygen atoms. The oxygen atoms shown by the arrowheads belong to the side chains of, from top, Ser replacing Phe617 of AcrB, Tyr replacing Ile277, and Thr replacing Phe611.

Natural Substrates for Multidrug Efflux RND Pumps

The idea that the multidrug efflux pumps may have other, more physiological functions than the extrusion of exogenous chemicals began with the finding that an MFS pump in B. subtilis, Blt, is actually a component of an operon that detoxifies spermidine. Clearly the RND-type multidrug pumps have not appeared in response to the widespread use of antibiotics, as most of these pumps are coded by chromosomal genes that are present in strains isolated long before the antibiotic era. For AcrB of E. coli, its main physiological function is to protect the bacteria against bile salts, detergents that are abundant in the environment where E. coli lives, the intestinal tract of vertebrates. This notion is supported by the observation that bile salts appear to have the highest affinity, among substrates, to the AcrB transporter. The AcrAB-TolC system was reported to be important in the pathogenicity of Salmonella spp. However, this conclusion was reached by orally infecting experimental animals with bacteria and seems less convincing because the infecting bacteria will obviously encounter bile acids where the AcrAB pump will be needed for their survival. With plant pathogens, again RND pumps may be needed to protect the bacteria against plant secondary metabolites. In Pseudomonas syringae, the inactivation of one RND pump was reported to decrease the secretion of lipopeptide phytotoxins, but the decrease was only modest (41% to 67%).

With soil organisms such as P. aeruginosa, it seems likely that most of the RND pumps are involved in excluding toxic compounds produced by other soil microorganisms. The reports that they are involved in the export of homoserine lactone autoinducers are not convincing, as these compounds should diffuse across membranes spontaneously, and as the assay involves cells that are secreting these compounds and at the same time are exposed to the same compound secreted by other cells. However, the loss of some RND pumps was shown to result in the loss of invasiveness in the cell culture, through mechanisms that are unclear.

Contributions of RND Efflux Pumps to the Resistance in Clinical Strains

RND pumps are making major contributions both to the intrinsic and elevated resistance of clinically relevant pathogens. For example, although fluoroquinoline resistance is often attributed to the mutational alteration of the target topoisomerases, high-level resistance seems to require an added contribution by increased efflux. Fluoroquinolone resistance in P. aeruginosa is rapidly increasing; efflux makes a major contribution in most such cases. A significant portion of aminoglycoside-resistant clinical isolates of gram-negative bacteria was previously classified as being caused by "decreased permeability." These strains are now known to owe their resistance to increased active efflux. In view of the important roles multidrug efflux plays in resistance, many scientists are concerned with the trend to increasingly add disinfectants to household products like soaps, because such compounds may select for pump overproduction mutants.

RND Transporters that Pump out Nondrug Substrates

Some of the RND transporters are involved in the export of compounds other than the antimicrobials. The "Nodulation" part of its name came from the putative NodGHI proteins, which are now fused into one RND protein, involved in the secretion of "nodulation factors" or lipochitin oligosaccharides by Rhizobium.

Typical multidrug exporters such as AcrB and MexB also export some simple solvents. Organisms that resist high concentrations of solvents, such as Pseudomonas putida, contain RND transporters optimized for the efflux of solvents, some of which can also pump out antimicrobials. Among the human proteins, the Niemann-Pick C1 disease protein, a member of RND family, attracted attention because mutations in the protein cause cholesterol accumulation in some organelles. Indeed, a close relative of this protein, Niemann-Pick C1-like protein, is involved in cholesterol transport in the intestinal epithelial cells.

There are many examples of RND transporters in the genome of M. tuberculosis. One of them (MmpL7) is involved in the export of complex, apolar lipid, phthiocerol dimycocerosate and another one (MmpL8) in that of 2,3-diacyl-α,α′-trehalose-2′-sulfate, a precursor of sulfatides.

RND transporters also play a major role in resistance of gram-negative bacteria against toxic metals. CzrA, an RND transporter in Ralstonia that pumps out Co^{2+}, Ni^{2+}, Cd^{2+}, and Zn^{2+}, has been studied intensively, and an early successful reconstitution of an RND pump has been achieved in this system.

Other Multidrug Efflux Pumps Energized by Ionic Gradients

Several multidrug efflux transporters belong to the Multidrug and Toxin Extrusion (MATE) family. NorM of Vibrio parahaemolyticus, a protein with 12 transmembrane helices, pumps out fluoroquinolones and ethidium in exchange for the influx of Na+ (141). A similar pump was found in Bacteroides fragilis.

Multidrug Efflux Pumps of the ATP-binding Cassette Superfamily

Transporters of ATP-Binding Cassette (ABC) superfamily play a major role in the multidrug resistance of cancer cells. Their role in drug resistance, however, seems to be more limited in bacteria. LmrA of L. lactis is homologous to one-half of the mammalian MDR1 protein. Although its overproduction in E. coli confers resistance to cationic dyes, daunomycin, and triphenylphosphonium, it does not seem to play a significant role in the drug resistance of L. lactis. Nevertheless, biochemical studies of this pump led to a proposed mechanism for the coupling of drug extrusion and ATP hydrolysis and to the discovery that the transmembrane domain of this pump can alone catalyze a proton-gradient-dependent flux of drugs. Studies were made on other ABC drug exporters of gram-positive bacteria, such as BmrA of B. subtilis and Sav1866 of S. aureus, whose X-ray crystallographic structure had a major impact in the studies of ABC transporters.

Another ABC transporter, MacB in E. coli, occurs together with the structural gene for a periplasmic adaptor protein MacA and confers resistance to macrolides when overexpressed. In vitro reconstitution showed that the ATPase activity of MacB was stimulated by MacA, showing that the membrane fusion protein has functions other than the structural one.

Multidrug Resistance caused by Altered Physiological States

The antibiotic susceptibility of bacterial cells is affected by their physiological states. One important consequence of this phenomenon is the occurrence of "persister" cells. Thus, it was discovered early that even high concentrations of antibiotics do not kill all of the bacterial population, leaving behind a persister population that is genetically identical with the susceptible cells. When biofilms were found more resistant to most antibiotics, there were initial attempts to interpret this finding on the basis of a more limited diffusion of drugs through the biofilm structure. However, this mechanism cannot produce large increases in resistance. Although there are interesting data that link drug resistance in P. aeruginosa biofilms to the production of periplasmic β-(1–3)-glucans and an efflux system, it is difficult, at least at present, to explain the extensive antibiotic resistance of biofilms on the basis of alteration of all cells in the population.

It is therefore attractive to explain such resistance as the result of the presence of a large number of persister cells in the biofilm population. The presence of persisters is now thought to be an example of the strategy whereby bacteria naturally generate mixtures of phenotypically different populations, so that one of them can be advantageous to a changing environmental demand. Persisters limit the efficacy of antibiotic therapy, and we note that a recent single-cell study has identified an antibiotic-susceptible phase in the life cycle of typical persister cells.

Multidrug-resistant Gram-negative Bacteria

Multidrug resistant Gram-negative bacteria (MDRGN bacteria) are a type of Gram-negative bacteria with resistance to multiple antibiotics. They can cause bacteria infections that pose a serious and rapidly emerging threat for hospitalized patients and especially patients in intensive care units. Infections caused by MDR strains are correlated with increased morbidity, mortality, and prolonged hospitalization. Thus, not only do these bacteria pose a threat to global public health, but also create a significant burden to healthcare systems.

Emerging Threat

These bacteria pose a great threat to public health due to the limited treatment options available as well as lack of newly developed antimicrobial medications. MDR strains of *Enterobacteriaceae*, *Pseudomonas aeruginosa*, and *Acinetobacter baumannii* have become of most concern because they have been reported by hospitals all around the United States. There are many factors which could be contributed to the existence and spread of MDR gram-negative bacteria such as the: overuse of existing antimicrobial agents, which has led to the development of adaptive resistance mechanisms by bacteria; a lack of responsible antimicrobial stewardship such that the use of multiple broad-spectrum agents has helped perpetuate the cycle of increasing resistance; and a lack of good infection control practices.

Treatment Options

Although there is currently a shortage of new drugs in the antimicrobial realm, there are a few antibiotics currently being studied and tested for the treatment of serious Gram-negative bacterial infections. These include cephalosporins, ceftobiprole, ceftarolin and FR-264205. The lack of newly emerging antimicobrial drugs have resulted in the revisit of old antibiotic drugs such as colistin and fosfomycin (Polymyxins), which are traditionally considered to be toxic but have gained a principal role in the treatment of the most problematic MDR Gram-negative pathogens including *Pseudomonas aeruginosa, Acinetobacter baumannii, Klebsiella pneumoniae* and *Stenotrophomonas maltophilia*. Also, there has been interest in the drug Tigecycline, which is from the class of antibiotics called glycylcyclines, for treating MDRGN infections. This drug shows promise in infections from multi-drug resistant *K. pneumoniae* (K. pneumoniae carbapenemase [KPC]- and ESBL-producing strains) and *Enterobacteriaceae* with various mechanisms of resistance. A review of investigational antibiotics shows that several new agents will become available in the coming years, even though the pace of antimicrobial research has proven far too slow:

> "Overuse of antimicrobial agents and problems with infection control practices have led to the development of multidrug-resistant gram-negative bacterial infections. We used to use carbapenems as the main option in several countries for those severe infections; however, now there are several mechanisms of resistance, including carbapenemase production among *Acinetobacter*, *Pseudomonas*, and *Klebsiella* isolates. Subsequently, carbapenems are sometimes not active against those serious infections. That is why clinicians around the world have reconsidered the use of older antimicrobial agents, including polymyxins. — Matthew Falagas, MD."

A study of MDRGN in long-term care facilities, reported in 2010, concluded that patients with severe dementia who require assistance with the activities of daily life are at high risk of MDRGN co-colonization and may be the "superspreaders" of MDRGN in these facilities.

Vancomycin-resistant Enterococcus

Vancomycin-resistant *Enterococcus*, or vancomycin-resistant enterococci (VRE), are bacterial strains of the genus *Enterococcus* that are resistant to the antibiotic vancomycin.

Mechanism of Acquired Resistance

Six different types of vancomycin resistance are shown by enterococcus: Van-A, Van-B, Van-C, Van-D, Van-E and Van-G. The significance is that Van-A VRE is resistant to both vancomycin and teicoplanin, Van-B VRE is resistant to vancomycin but susceptible to teicoplanin, and Van-C is only partly resistant to vancomycin

The mechanism of resistance to vancomycin found in enterococcus involves the alteration of the peptidoglycan synthesis pathway. The D-alanyl-D-lactate variation results in the loss of one hydrogen-bonding interaction (four, as opposed to five for D-alanyl-D-alanine) being possible between vancomycin and the peptide. The D-alanyl-D-serine variation causes a six-fold loss of affinity between vancomycin and the peptide, likely due to steric hindrance.

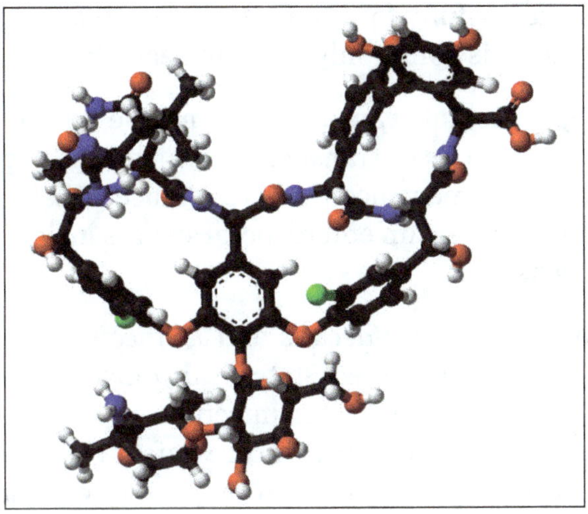

Vancomycin.

Screening

Screening for VRE can be accomplished in a number of ways. For inoculating peri-rectal/anal swabs or stool specimens directly, one method uses bile esculin azide agar plates containing 6 μg/ml of vancomycin. Black colonies should be identified as an enterococcus to species level and further confirmed as vancomycin resistant by an MIC method before reporting as VRE.

Vancomycin resistance can be determined for enterococcal colonies available in pure culture by inoculating a suspension of the organism onto a commercially available brain heart infusion agar (BHIA) plate containing 6 μg/ml vancomycin. The National Committee for Clinical Laboratory Standards (NCCLS) recommends performing a vancomycin MIC test and also motility and pigment production tests to distinguish species with acquired resistance (vanA and vanB) from those with vanC intrinsic resistance.

Treatment of Infection

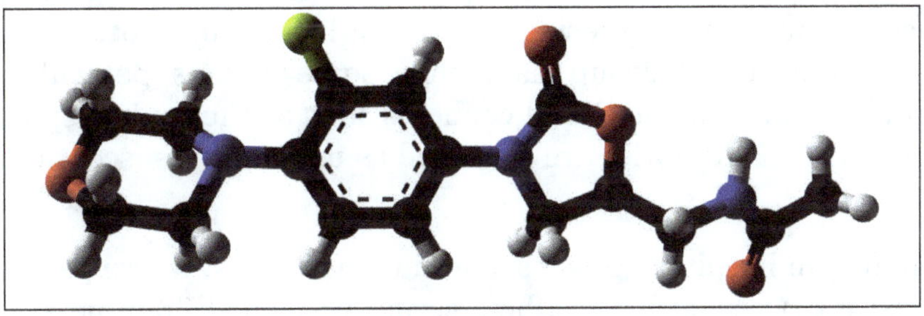

Linezolid.

Cephalosporin use is a risk factor for colonization and infection by VRE, and restriction of cephalosporin usage has been associated with decreased VRE infection and transmission in hospitals. *Lactobacillus rhamnosus* GG (LGG), a strain of *L.*

rhamnosus, was used successfully for the first time to treat gastrointestinal carriage of VRE. In the US, linezolid is commonly used to treat VRE.

To become vancomycin-resistant, vancomycin-sensitive enterococci typically obtain new DNA in the form of plasmids or transposons which encode genes that confer vancomycin resistance. This acquired vancomycin resistance is distinguished from the natural vancomycin resistance of certain enterococcal species including *E. gallinarum* and *E. casseliflavus/flavescens*.

High-level vancomycin-resistant *E. faecalis* and *E. faecium* are clinical isolates first documented in the 1980s. In the United States, vancomycin-resistant *E. faecium* was associated with 4% of healthcare-associated infections reported to the Centers for Disease Control and Prevention National Healthcare Safety Network from January 2006 to October 2007. VRE can be carried by healthy people who have come into contact with the bacteria, usually in a hospital (nosocomial infection), although it is thought that a significant percentage of intensively farmed chicken also carry VRE. Other regions have noted a similar distribution, but with increased incidence of VRE. For example, a 2006 study of nosocomial VRE revealed a rapid spread of resistance among enterococci along with an emerging shift in VRE distribution in the Middle East region, such as Iran. Treatment failures in enterococcal infections result from inadequate information regarding glycopeptide resistance of endemic enterococci due to factors such as the presence of VanA and VanB. The study from Iran reported the first case of VRE isolates that carried VanB gene in enterococcal strains from Iran. This study also noted the first documented isolation of nosocomial *E. raffinosus* and *E. mundtii* in the Middle East region.

Methicillin-resistant Staphylococcus Aureus

Methicillin-resistant *Staphylococcus aureus* (MRSA) refers to a group of Gram-positive bacteria that are genetically distinct from other strains of *Staphylococcus aureus*. MRSA is responsible for several difficult-to-treat infections in humans. MRSA is any strain of *S. aureus* that has developed, through horizontal gene transfer and natural selection, multiple drug resistance to beta-lactam antibiotics. β-lactam antibiotics are a broad-spectrum group that include some penams (penicillin derivatives such as methicillin and oxacillin) and cephems such as the cephalosporins. Strains unable to resist these antibiotics are classified as methicillin-susceptible *S. aureus*, or MSSA.

MRSA is common in hospitals, prisons, and nursing homes, where people with open wounds, invasive devices such as catheters, and weakened immune systems are at greater risk of hospital-acquired infection. MRSA began as a hospital-acquired infection, but has become community-acquired, as well as livestock-acquired. The terms HA-MRSA (healthcare-associated or hospital-acquired MRSA), CA-MRSA (community-associated MRSA), and LA-MRSA (livestock-associated MRSA) reflect this.

Signs and Symptoms

In humans, *Staphylococcus aureus* is part of the normal microbiota present in the upper respiratory tract, and on skin and in the gut mucosa. *S. aureus*, along with similar bacterial species that can colonize and act symbiotically but can cause disease if they begin to take over the tissues they have colonized or invade other tissues, has been called a "pathobiont".

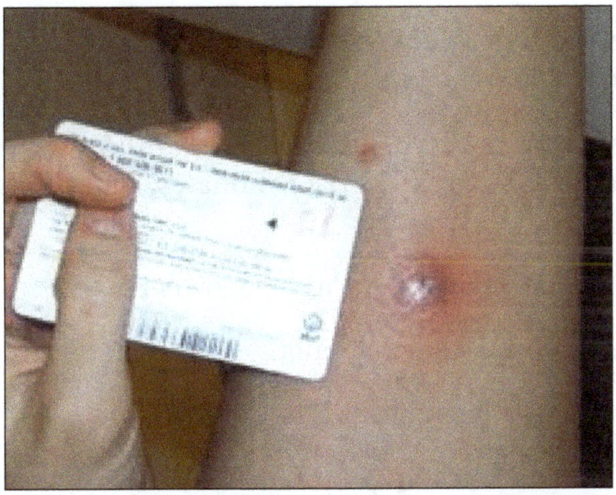

Although usually carried without symptoms, MRSA often presents as small red pustular skin infections

After 72 hours, MRSA can take hold in human tissues and eventually become resistant to treatment. The initial presentation of MRSA is small red bumps that resemble pimples, spider bites, or boils; they may be accompanied by fever and, occasionally, rashes. Within a few days, the bumps become larger and more painful; they eventually open into deep, pus-filled boils. About 75 percent of CA-MRSA infections are localized to skin and soft tissue and usually can be treated effectively.

Risk Factors

A select few of the populations at risk include:

- People with indwelling implants, prostheses, drains, and catheters.
- People who are frequently in crowded places, especially with shared equipment and skin-to-skin contact.
- People with weak immune systems (HIV/AIDS, lupus, or cancer sufferers; transplant recipients; severe asthmatics; etc.)
- Diabetics.
- Intravenous drug users.
- Users of quinolone antibiotics.

- Elderly people.
- School children sharing sports and other equipment.
- College students living in dormitories.
- People staying or working in a health-care facility for an extended period of time.
- People who spend time in coastal waters where MRSA is present, such as some beaches in Florida and the West Coast of the United States.
- People who spend time in confined spaces with other people, including occupants of homeless shelters, prison inmates, and military recruits in basic training.
- Veterinarians, livestock handlers, and pet owners.
- People who ingest unpasteurized milk.
- People who are immunocompromised and also colonized.
- People with chronic obstructive pulmonary disease.
- People who have had thoracic surgery.

As many as 22% of people infected with MRSA do not have any discernable risk factors.

Hospitalized People

People who are hospitalized, including the elderly, are often immunocompromised and susceptible to infection of all kinds, including MRSA; an infection by MRSA is called healthcare-associated or hospital-acquired methicillin-resistant *S. aureus* (HA-MRSA). Generally, those infected by MRSA stay infected for just under 10 days, if treated by a doctor, although effects may vary from person to person.

Both surgical and nonsurgical wounds can be infected with HA-MRSA. Surgical site infections occur on the skin surface, but can spread to internal organs and blood to cause sepsis. Transmission can occur between healthcare providers and patients because some providers may neglect to perform preventative hand-washing between examinations.

People in nursing homes are at risk for all the reasons above, further complicated by their generally weaker immune systems.

Prison Inmates and Military Recruits

Prisons and military barracks can be crowded and confined, and poor hygiene practices may proliferate, thus putting inhabitants at increased risk of contracting MRSA. Cases of MRSA in such populations were first reported in the United States and later in Canada. The earliest reports were made by the Centers for Disease Control and Prevention in US state prisons. In the news media, hundreds of reports of MRSA outbreaks in prisons appeared between 2000 and 2008. For example, in February 2008, the Tulsa County jail in Oklahoma started treating an average of 12 *S. aureus* cases per month.

Animals

Antibiotic use in livestock increases the risk that MRSA will develop among the livestock; strains MRSA ST 398 and CC398 are transmissible to humans. Generally, animals are asymptomatic.

Domestic pets are susceptible to MRSA infection by transmission from their owners; conversely, MRSA-infected pets can also transmit MRSA to humans.

Athletes

Locker rooms, gyms, and related athletic facilities offer potential sites for MRSA contamination and infection. Athletes have been identified as a high-risk group. A study linked MRSA to the abrasions caused by artificial turf. Three studies by the Texas State Department of Health found the infection rate among football players was 16 times the national average. In October 2006, a high-school football player was temporarily paralyzed from MRSA-infected turf burns. His infection returned in January 2007 and required three surgeries to remove infected tissue, and three weeks of hospital stay.

In 2013, Lawrence Tynes, Carl Nicks, and Johnthan Banks of the Tampa Bay Buccaneers were diagnosed with MRSA. Tynes and Nicks apparently did not contract the infection from each other, but whether Banks contracted it from either individual is unknown. In 2015, Los Angeles Dodgers infielder Justin Turner was infected while the team visited the New York Mets. In October 2015, New York Giants tight end Daniel Fells was hospitalized with a serious MRSA infection.

Children

MRSA is becoming a critical problem in children; studies found 4.6% of patients in U.S. health-care facilities, (presumably) including hospital nurseries, were infected or colonized with MRSA. Children and adults who come in contact with daycare centers, playgrounds, locker rooms, camps, dormitories, classrooms and other school settings, and gyms and workout facilities are at higher risk of contracting MRSA. Parents should be especially cautious of children who participate in activities where sports equipment is shared, such as football helmets and uniforms.

Mechanism

Antimicrobial resistance is genetically based; resistance is mediated by the acquisition of extrachromosomal genetic elements containing genes that confer resistance to certain antibiotics. Examples of such elements include plasmids, transposable genetic elements, and genomic islands, which can be transferred between bacteria through horizontal gene transfer. A defining characteristic of MRSA is its ability to thrive in the presence of penicillin-like antibiotics, which normally prevent bacterial growth by inhibiting synthesis of cell wall material. This is due to a resistance gene, *mecA*, which stops β-lactam antibiotics from inactivating the enzymes (transpeptidases) critical for cell wall synthesis.

SCCmec

Staphylococcal cassette chromosome *mec* (SCC*mec*) is a genomic island of unknown origin containing the antibiotic resistance gene *mecA*. SCC*mec* contains additional genes beyond *mecA*, including the cytolysin gene *psm-mec*, which may suppress virulence in HA-acquired MRSA strains. In addition, this locus encodes strain-dependent gene regulatory RNAs known as *psm-mec*RNA. SCC*mec* also contains *ccrA* and *ccrB*; both genes encode recombinases that mediate the site-specific integration and excision of the SCC*mec* element from the *S. aureus* chromosome. Currently, six unique SCC*mec* types ranging in size from 21–67 kb have been identified; they are designated types I–VI and are distinguished by variation in *mec* and *ccr* gene complexes. Owing to the size of the SCC*mec* element and the constraints of horizontal gene transfer, a minimum of five clones are thought to be responsible for the spread of MRSA infections, with clonal complex (CC) 8 most prevalent. SCC*mec* is thought to have originated in the closely related *Staphylococcus sciuri* species and transferred horizontally to *S. aureus*.

Different SCC*mec* genotypes confer different microbiological characteristics, such as different antimicrobial resistance rates. Different genotypes are also associated with different types of infections. Types I–III SCC*mec* are large elements that typically contain additional resistance genes and are characteristically isolated from HA-MRSA strains. Conversely, CA-MRSA is associated with types IV and V, which are smaller and lack resistance genes other than *mecA*.

These distinctions were thoroughly investigated by Collins et al. in 2001, and can be explained by the fitness differences associated with carriage of a large or small SCC*mec* plasmid. Carriage of large plasmids, such as SCC*mec*I–III, is costly to the bacteria, resulting in a compensatory decrease in virulence expression. MRSA is able to thrive in hospital settings with increased antibiotic resistance but decreased virulence – HA-MRSA targets immunocompromised, hospitalized hosts, thus a decrease in virulence is not maladaptive. In contrast, CA-MRSA tends to carry lower-fitness cost SCC*mec* elements to offset the increased virulence and toxicity expression required to infect healthy hosts.

mecA

mecA is a biomarker gene responsible for resistance to methicillin and other β-lactam antibiotics. After acquisition of *mecA*, the gene must be integrated and localized in the *S. aureus* chromosome. *mecA* encodes penicillin-binding protein 2a (PBP2a), which differs from other penicillin-binding proteins as its active site does not bind methicillin or other β-lactam antibiotics. As such, PBP2a can continue to catalyze the transpeptidation reaction required for peptidoglycan cross-linking, enabling cell wall synthesis even in the presence of antibiotics. As a consequence of the inability of PBP2a to interact with β-lactam moieties, acquisition of *mecA* confers resistance to all β-lactam antibiotics in addition to methicillin.

mecA is under the control of two regulatory genes, *mecI* and *mecR1*. MecI is usually bound to the *mecA* promoter and functions as a repressor. In the presence of a β-lactam antibiotic, MecR1 initiates a signal transduction cascade that leads to transcriptional activation of *mecA*. This is achieved by MecR1-mediated cleavage of MecI, which alleviates MecI repression. *mecA* is further controlled by two co-repressors, *blaI* and *blaR1*. *blaI* and *blaR1* are homologous to *mecI* and *mecR1*, respectively, and normally function as regulators of *blaZ*, which is responsible for penicillin resistance. The DNA sequences bound by *mecI* and *blaI* are identical; therefore, *blaI* can also bind the *mecA* operator to repress transcription of *mecA*.

Arginine Catabolic Mobile Element

The arginine catabolic mobile element (ACME) is a virulence factor present in many MRSA strains but not prevalent in MSSA. SpeG-positive ACME compensates for the polyamine hypersensitivity of *S. aureus* and facilitates stable skin colonization, wound infection, and person-to-person transmission.

Strains

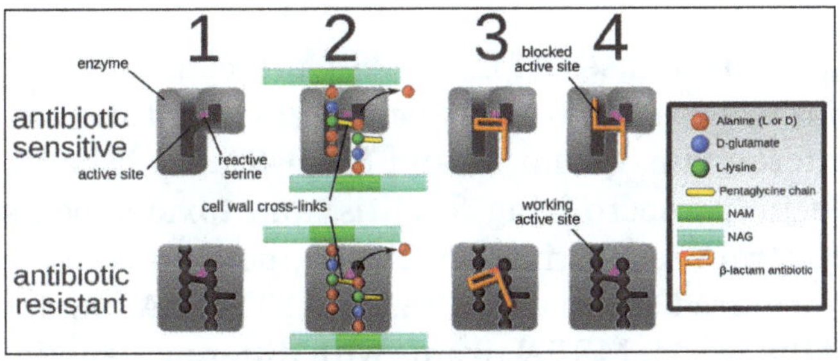

Diagram depicting antibiotic resistance through alteration of the antibiotic's target site, modeled after MRSA's resistance to penicillin. Beta-lactam antibiotics permanently inactivate PBP enzymes, which are essential for cell wall synthesis and thus for bacterial life, by permanently binding to their active sites. Some forms of MRSA, however, express a different PBP that will not allow the antibiotic into its active site.

Acquisition of SCC*mec* in methicillin-sensitive *S. aureus* (MSSA) gives rise to a number of genetically different MRSA lineages. These genetic variations within different MRSA strains possibly explain the variability in virulence and associated MRSA infections.

The first MRSA strain, ST250 MRSA-1, originated from SCC*mec* and ST250-MSSA integration. Historically, major MRSA clones ST2470-MRSA-I, ST239-MRSA-III, ST5-MRSA-II, and ST5-MRSA-IV were responsible for causing hospital-acquired MRSA (HA-MRSA) infections. ST239-MRSA-III, known as the Brazilian clone, was highly transmissible compared to others and distributed in Argentina, Czech Republic, and Portugal.

In the UK, the most common strains of MRSA are EMRSA15 and EMRSA16. EMRSA16 has been found to be identical to the ST36:USA200 strain, which circulates in the United States, and to carry the SCC*mec* type II, enterotoxin A and toxic shock syndrome toxin 1 genes. Under the new international typing system, this strain is now called MRSA252. EMRSA 15 is also found to be one of the common MRSA strains in Asia. Other common strains include ST5:USA100 and EMRSA 1. These strains are genetic characteristics of HA-MRSA.

Community-acquired MRSA (CA-MRSA) strains emerged in late 1990 to 2000, infecting healthy people who had not been in contact with healthcare facilities. Researchers suggest that CA-MRSA did not evolve from HA-MRSA. This is further proven by molecular typing of CA-MRSA strains and genome comparison between CA-MRSA and HA-MRSA, which indicate that novel MRSA strains integrated SCC*mec* into MSSA separately on its own. By mid-2000, CA-MRSA was introduced into healthcare systems and distinguishing CA-MRSA from HA-MRSA became a difficult process. Community-acquired MRSA is more easily treated and more virulent than hospital-acquired MRSA (HA-MRSA). The genetic mechanism for the enhanced virulence in CA-MRSA remains an active area of research. The Panton–Valentine leukocidin (PVL) genes are of particular interest because they are a unique feature of CA-MRSA.

In the United States, most cases of CA-MRSA are caused by a CC8 strain designated ST8:USA300, which carries SCC*mec* type IV, Panton–Valentine leukocidin, PSM-alpha and enterotoxins Q and K, and ST1:USA400. The ST8:USA300 strain results in skin infections, necrotizing fasciitis, and toxic shock syndrome, whereas the ST1:USA400 strain results in necrotizing pneumonia and pulmonary sepsis. Other community-acquired strains of MRSA are ST8:USA500 and ST59:USA1000. In many nations of the world, MRSA strains with different genetic background types have come to predominate among CA-MRSA strains; USA300 easily tops the list in the U.S. and is becoming more common in Canada after its first appearance there in 2004. For example, in Australia, ST93 strains are common, while in continental Europe ST80 strains, which carry SCC*mec* type IV, predominate. In Taiwan, ST59 strains, some of which are resistant to many non-beta-lactam antibiotics, have arisen as common causes of skin and soft tissue infections in the community. In a remote region of Alaska, unlike most of the continental U.S., USA300 was found only rarely in a study of MRSA strains from outbreaks in 1996 and 2000 as well as in surveillance from 2004–06.

A MRSA strain, CC398, is found in intensively reared production animals (primarily pigs, but also cattle and poultry), where it can be transmitted to humans as LA-MRSA (livestock-associated MRSA).

Diagnosis

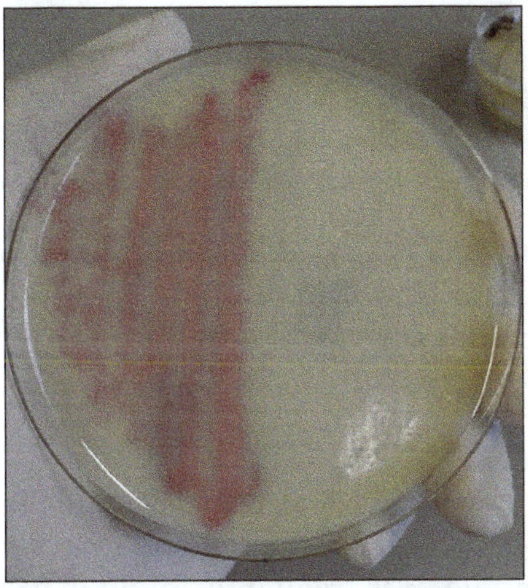

A selective and differential chromogenic medium for the qualitative direct detection of MRSA.

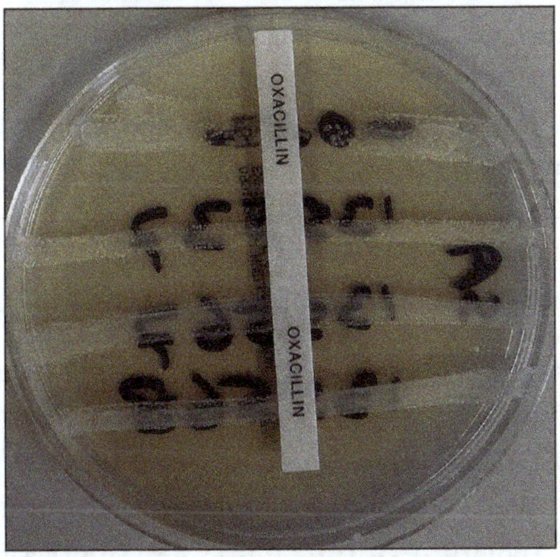

MRSA resistance to oxacillin being tested. The top *S. aureus* isolate is a control that is not resistant to oxacillin; the other three isolates are MRSA-positive.

Diagnostic microbiology laboratories and reference laboratories are key for identifying outbreaks of MRSA. Normally, a bacterium must be cultured from blood, urine, sputum, or other body-fluid samples, and in sufficient quantities to perform confirmatory tests early-on. Still, because no quick and easy method exists to diagnose MRSA, initial treatment of the infection is often based upon "strong suspicion" and techniques by the treating physician; these include quantitative PCR procedures, which are employed in clinical laboratories for quickly detecting and identifying MRSA strains.

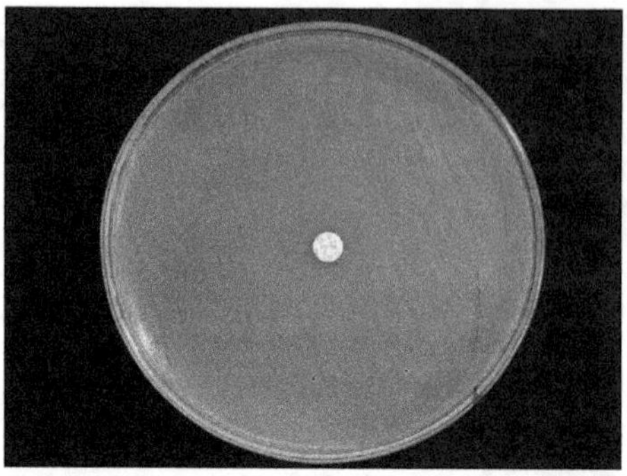

Mueller-Hinton agar showing MRSA resistant to an oxacillin disk.

Another common laboratory test is a rapid latex agglutination test that detects the PB-P2a protein. PBP2a is a variant penicillin-binding protein that imparts the ability of *S. aureus* to be resistant to oxacillin.

Microbiology

Like all *S. aureus* (also abbreviated SA at times), methicillin-resistant *S. aureus* is a Gram-positive, spherical (coccus) bacterium about 1 micron in diameter. It does not form spores and it is not motile. It is frequently found in grape-like clusters or chains. Unlike methicillin-susceptible *S. aureus* (MSSA), MRSA is slow-growing on a variety of media and has been found to exist in mixed colonies of MSSA. The *mecA* gene, which confers resistance to a number of antibiotics, is always present in MRSA and usually absent in MSSA; however, in some instances, the *mecA* gene is present in MSSA but is not expressed. Polymerase chain reaction (PCR) testing is the most precise method for identifying MRSA strains. Specialized culture media have been developed to better differentiate between MSSA and MRSA and, in some cases, such media can be used to identify specific strains that are resistant to different antibiotics.

Other strains of *S. aureus* have emerged that are resistant to oxacillin, clindamycin, teicoplanin, and erythromycin. These resistant strains may or may not possess the *mecA* gene. *S. aureus* has also developed resistance to vancomycin (VRSA). One strain is only partially susceptible to vancomycin and is called vancomycin-intermediate *S. aureus* (VISA). GISA, a strain of resistant *S. aureus*, is glycopeptide-intermediate *S. aureus* and is less suspectible to vancomycin and teicoplanin. Resistance to antibiotics in *S. aureus* can be quantified by determining the amount of the antibiotic that must be used to inhibit growth. If *S. aureus* is inhibited at a concentration of vancomycin less than or equal to 4 μg/ml, it is said to be susceptible. If a concentration greater than 32 μg/ml is necessary to inhibit growth, it is said to be resistant.

Prevention

Screening

In health-care settings, isolating those with MRSA from those without the infection is one method to prevent transmission. Rapid culture and sensitivity testing and molecular testing identifies carriers and reduces infection rates.

MRSA can be identified by swabbing the nostrils and isolating the bacteria found there. Combined with extra sanitary measures for those in contact with infected people, swab screening people admitted to hospitals has been found to be effective in minimizing the spread of MRSA in hospitals in the United States, Denmark, Finland, and the Netherlands.

Handwashing

The Centers for Disease Control and Prevention offers suggestions for preventing the contraction and spread of MRSA infection which are applicable to those in community settings, including incarcerated populations, childcare center employees, and athletes. To prevent the spread of MRSA, the recommendations are to wash hands thoroughly and regularly using soap and water or an alcohol-based sanitizer. Additional recommendations are to keep wounds clean and covered, avoid contact with other people's wounds, avoid sharing personal items such as razors or towels, shower after exercising at athletic facilities, and shower before using swimming pools or whirlpools.

Isolation

Excluding medical facilities, current US guidance does not require workers with MRSA infections to be routinely excluded from the general workplace. The National Institutes of Health recommend that those with wound drainage that cannot be covered and contained with a clean, dry bandage and those who cannot maintain good hygiene practices be reassigned. Workers with active infections are excluded from activities where skin-to-skin contact is likely to occur. To prevent the spread of staphylococci or MRSA in the workplace, employers are encouraged to make available adequate facilities that support good hygiene. In addition, surface and equipment sanitizing should conform to Environmental Protection Agency-registered disinfectants. In hospital settings, contact isolation can be stopped after one to three cultures come back negative.

To prevent the spread of MRSA in the home, health departments recommend laundering materials that have come into contact with infected persons separately and with a dilute bleach solution; to reduce the bacterial load in one's nose and skin; and to clean and disinfect those things in the house that people regularly touch, such as sinks, tubs, kitchen counters, cell phones, light switches, doorknobs, phones, toilets, and computer keyboards.

Restricting Antibiotic Use

Glycopeptides, cephalosporins, and in particular, quinolones are associated with an increased risk of colonisation of MRSA. Reducing use of antibiotic classes that promote MRSA colonisation, especially fluoroquinolones, is recommended in current guidelines.

Public Health Considerations

Mathematical models describe one way in which a loss of infection control can occur after measures for screening and isolation seem to be effective for years, as happened in the UK. In the "search and destroy" strategy that was employed by all UK hospitals until the mid-1990s, all hospitalized people with MRSA were immediately isolated, and all staff were screened for MRSA and were prevented from working until they had completed a course of eradication therapy that was proven to work. Loss of control occurs because colonised people are discharged back into the community and then readmitted; when the number of colonised people in the community reaches a certain threshold, the "search and destroy" strategy is overwhelmed. One of the few countries not to have been overwhelmed by MRSA is the Netherlands: an important part of the success of the Dutch strategy may have been to attempt eradication of carriage upon discharge from hospital.

Decolonization

As of 2013, no randomized clinical trials had been conducted to understand how to treat nonsurgical wounds that had been colonized, but not infected, with MRSA, and insufficient studies had been conducted to understand how to treat surgical wounds that had been colonized with MRSA. As of 2013, whether strategies to eradicate MRSA colonization of people in nursing homes reduced infection rates was not known.

Care should be taken when trying to drain boils, as disruption of surrounding tissue can lead to larger infections, or even infection of the blood stream (often with fatal consequences). Mupirocin 2% ointment can be effective at reducing the size of lesions. A secondary covering of clothing is preferred. As shown in an animal study with diabetic mice, the topical application of a mixture of sugar (70%) and 3% povidone-iodine paste is an effective agent for the treatment of diabetic ulcers with MRSA infection.

Community Settings

Maintaining the necessary cleanliness may be difficult for people if they do not have access to facilities such as public toilets with handwashing facilities. In the United Kingdom, the Workplace (Health, Safety and Welfare) Regulations 1992 require businesses to provide toilets for their employees, along with washing facilities including soap or other suitable means of cleaning. Guidance on how many toilets to provide and what sort of washing facilities should be provided alongside them is given in the Workplace (Health, Safety and Welfare) Approved Code of Practice and Guidance L24, available from Health and Safety Executive Books, but no legal obligations exist on local authorities in the United Kingdom to provide public toilets, and although in 2008, the House of Commons Communities and Local Government Committee called for a duty on local authorities to develop a public toilet strategy, this was rejected by the Government.

Agriculture

Some advocate regulations on the use of antibiotics in animal feed to prevent the emergence of drug-resistant strains of MRSA. MRSA is established in animals and birds.

Treatment

Antibiotics

Treatment of MRSA infection is urgent and delays can be fatal. The location and history related to the infection determines the treatment. The route of administration of an antibiotic varies. Antibiotics effective against MRSA can be given by IV, oral, or a combination of both, and depend on the specific circumstances and patient characteristics. The use of concurrent treatment with vancomycin or other beta-lactam agents may have a synergistic effect.

Both CA-MRSA and HA-MRSA are resistant to traditional anti-staphylococcal beta-lactam antibiotics, such as cephalexin. CA-MRSA has a greater spectrum of antimicrobial susceptibility to sulfa drugs (like co-trimoxazole (trimethoprim/sulfamethoxazole), tetracyclines (like doxycycline and minocycline) and clindamycin (for osteomyelitis). MRSA can be eradicated with a regimen of linezolid, though treatment protocols vary and serum levels of antibiotics vary widely from person to person and may affect outcomes. The effective treatment of MRSA with linezolid has been successful in 87% of people. Linezolid is more effective in soft tissue infections than vancomycin. This is compared to eradication of infection in those with MRSA treated with vancomycin. Treatment with vancomycin is successful in approximately 49% of people. Linezolid belongs to the newer oxazolidinone class of antibiotics which has been shown to be effective against both CA-MRSA and HA-MRSA. The Infectious Disease Society of America recommends vancomycin, linezolid, or clindamycin (if susceptible) for treating those with MRSA pneumonia. Ceftaroline, a fifth-generation cephalosporin, is the first beta-lactam antibiotic approved in the US to treat MRSA infections in skin and soft tissue or community-acquired pneumonia.

Vancomycin and teicoplanin are glycopeptide antibiotics used to treat MRSA infections. Teicoplanin is a structural congener of vancomycin that has a similar activity spectrum but a longer half-life. Because the oral absorption of vancomycin and teicoplanin is very low, these agents can be administered intravenously to control systemic infections. Treatment of MRSA infection with vancomycin can be complicated, due to its inconvenient route of administration. Moreover, the efficacy of vancomycin against MRSA is inferior to that of anti-staphylococcal beta-lactam antibiotics against methicillin-susceptible *S. aureus* (MSSA).

Several newly discovered strains of MRSA show antibiotic resistance even to vancomycin and teicoplanin. These new strains of the MRSA bacterium have been dubbed vancomycin intermediate-resistant *S. aureus* (VISA). Linezolid, quinupristin/dalfopristin, daptomycin, ceftaroline, and tigecycline are used to treat more severe infections that do not respond to glycopeptides such as vancomycin. Current guidelines recommend daptomycin for VISA bloodstream infections and endocarditis.

This left vancomycin as the only effective agent available at the time. However, strains with intermediate (4–8 μg/ml) levels of resistance, termed glycopeptide-intermediate *S. aureus* (GISA) or vancomycin-intermediate *S. aureus* (VISA), began appearing in the late 1990s. The first identified case was in Japan in 1996, and strains have since been found in hospitals in England, France, and the US. The first documented strain with complete (>16 μg/ml) resistance to vancomycin, termed vancomycin-resistant *S. aureus* (VRSA) appeared in the United States in 2002. However, in 2011, a variant of vancomycin was tested that binds to the lactate variation and also binds well to the original target, thus reinstating potent antimicrobial activity.

Oxazolidinones such as linezolid became available in the 1990s and are comparable to vancomycin in effectiveness against MRSA. Linezolid resistance in *S. aureus* was reported in 2001, but infection rates have been at consistently low levels. In the United Kingdom and Ireland, no resistance was found in staphylococci collected from bacteremia cases between 2001 and 2006.

Skin and Soft-tissue Infections

In skin abscesses, the primary treatment recommended is removal of dead tissue, incision, and drainage. More information is needed to determine the effectiveness of specific antibiotics therapy in surgical site infections (SSIs). Examples of soft-tissue infections from MRSA include ulcers, impetigo, abscesses, and SSIs. In surgical wounds, evidence is weak (high risk of bias) that linezolid may be better than vancomycin to eradicate MRSA SSIs.

MRSA colonization is also found in nonsurgical wounds such as traumatic wounds, burns, and chronic ulcers (i.e.: diabetic ulcer, pressure ulcer, arterial insufficiency ulcer, venous ulcer). No conclusive evidence has been found about the best antibiotic regimen to treat MRSA colonization.

Children

In skin infections and secondary infection sites, topical mupirocin is used successfully. For bacteremia and endocarditis, vancomycin or daptomycin is considered. For children with MRSA-infected bone or joints, treatment is individualized and long-term. Neonates can develop neonatal pustulosis as a result of topical infection with MRSA. Clindamycin is not approved for the treatment of MRSA infection, but it is still used in children for soft-tissue infections.

Endocarditis and Bacteremia

Evaluation for the replacement of a prosthetic valve is considered. Appropriate antibiotic therapy may be administered for up to six weeks. Four to six weeks of antibiotic treatment is often recommended, and is dependent upon the extent of MRSA infection.

Respiratory Infections

CA-MRSA in hospitalized patients pneumonia treatment begins before culture results. After the susceptibility to antibiotics is performed, the infection may be treated with vancomycin or linezolid for up to 21 days. If the pneumonia is complicated by the accumulation of pus in the pleural cavity surrounding the lungs, drainage may be done along with antibiotic therapy. People with cystic fibrosis may develop respiratory complications related to MRSA infection. The incidence of MRSA in those with cystic fibrosis increased during 2000 to 2015 by five times. Most of these infections were HA-MRSA. MRSA accounts for 26% of lung infections in those with cystic fibrosis.

Bone and Joint Infections

Cleaning the wound of dead tissue and draining abscesses is the first action to treat the MRSA infection. Administration of antibiotics is not standardized and is adapted by a case-by-case basis. Antibiotic therapy can last up to 3 months and sometimes even longer.

Infected Implants

MRSA infection can occur associated with implants and joint replacements. Recommendations on treatment are based upon the length of time the implant has been in place. In cases of a recent placement of a surgical implant or artificial joint, the device may be retained while antibiotic therapy continues. If the placement of the device has occurred over 3 weeks ago, the device may be removed. Antibiotic therapy is used in each instance sometimes long-term.

Central Nervous System

MRSA can infect the central nervous system and form brain abscess, subdural empyema, and spinal epidural abscess. Excision and drainage can be done along with antibiotic treatment. Septic thrombosis of cavernous or dural venous sinus can sometimes be a complication.

Other Infections

Treatment is not standardized for other instances of MRSA infection in a wide range of tissues. Treatment varies for MRSA infections related to: subperiosteal abscesses, necrotizing pneumonia, cellulitis, pyomyositis, necrotizing fasciitis, mediastinitis, myocardial, perinephric, hepatic, and splenic abscesses, septic thrombophlebitis, and severe ocular infections, including endophthalmitis. Pets can be reservoirs and pass on MRSA to people. In some cases, the infection can be symptomatic and the pet can suffer a MRSA infection. Health departments recommend that the pet be taken to the veterinarian if MRSA infections keep occurring in the people who have contact with the pet.

Epidemiology

Worldwide, an estimated 2 billion people carry some form of *S. aureus*; of these, up to 53 million (2.7% of carriers) are thought to carry MRSA.

HA-MRSA

In a US cohort study of 1,300 healthy children, 2.4% carried MRSA in their nose. Bacterial sepsis occurs with most (75%) of cases of invasive MRSA infection. In 2009, there were an estimated 463,017 hospitalizations due to MRSA, or a rate of 11.74 per 1,000 hospitalizations. Many of these infections are less serious, but the Centers for Disease Control and Prevention (CDC) estimate that there are 80,461 invasive MRSA infections and 11,285 deaths due to MRSA annually. In 2003, the cost for a hospitalization due to MRSA infection was USD $92,363; a hospital stay for MSSA was $52,791.

Infection after surgery is relatively uncommon, but occurs as much as 33% in specific types of surgeries. Infections of surgical sites range from 1% to 33%. MRSA sepsis that occurs within 30 days following a surgical infection has a 15–38% mortality rate; MRSA sepsis that occurs within one year has a mortality rate of around 55%. There may be increased mortality associated with cardiac surgery. There is a rate of 12.9% in those infected with MRSA while only 3% infected with other organisms. SSIs infected with MRSA had longer hospital stays than those who did not.

Globally, MRSA infection rates are dynamic and vary year to year. According to the 2006 SENTRY Antimicrobial Surveillance Program report, the incidence of MRSA bloodstream infections was 35.9 per cent in North America. MRSA blood infections in Latin America was 29%. European incidence was 22.8%. The rate of all MRSA infections in Europe ranged from 50% per cent in Portugal down to 0.8 per cent in Sweden. Overall MRSA infection rates varied in Latin America: Colombia and Venezuela combined had 3%, Mexico had 50%, Chile 38%, Brazil 29%, and Argentina 28%.

The Centers for Disease Control and Prevention (CDC) estimated that about 1.7 million nosocomial infections occurred in the United States in 2002, with 99,000 associated deaths. The estimated incidence is 4.5 nosocomial infections per 100 admissions, with direct costs (at 2004 prices) ranging from $10,500 (£5300, €8000 at 2006 rates) per case (for bloodstream, urinary tract, or respiratory infections in immunocompetent people) to $111,000 (£57,000, €85,000) per case for antibiotic-resistant infections in the bloodstream in people with transplants. With these numbers, conservative estimates of the total direct costs of nosocomial infections are above $17 billion. The reduction of such infections forms an important component of efforts to improve healthcare safety. MRSA alone was associated with 8% of nosocomial infections reported to the CDC National Healthcare Safety Network from January 2006 to October 2007.

The British National Audit Office estimated that the incidence of nosocomial infections in Europe ranges from 4% to 10% of all hospital admissions. As of early 2005, the number of deaths in the United Kingdom attributed to MRSA has been estimated by various sources to lie in the area of 3,000 per year.

In the United States, an estimated 95 million people carry *S. aureus* in their noses; of these, 2.5 million (2.6% of carriers) carry MRSA. A population review conducted in three U.S. communities showed the annual incidence of CA-MRSA during 2001–2002 to be 18–25.7/100,000; most CA-MRSA isolates were associated with clinically relevant infections, and 23% of people required hospitalization.

CA-MRSA

In a US cohort study of 1,300 healthy children, 2.4% carried MRSA in their noses. There are concerns that the presence of MRSA in the environment may allow resistance to be transferred to other bacteria through phages (viruses that infect bacteria). The source of MRSA could come from hospital waste, farm sewage, or other waste water.

LA-MRSA

Livestock associated MRSA (LA-MRSA) has been observed in Korea, Brazil, Switzerland, Malaysia, India, Great Britain, Denmark, and China.

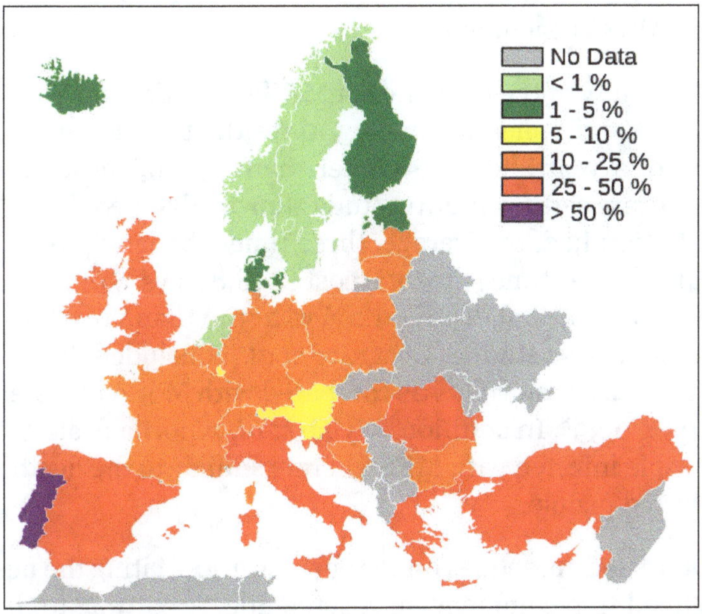

Incidence of MRSA in human blood samples in countries which took part in the study.

In 1961, the first known MRSA isolates were reported in a British study, and from 1961 to 1967, infrequent hospital outbreaks occurred in Western Europe and Australia, with methicillin then being licensed in England to treat resistant infections. Other reports of MRSA began to be described in the 1970s. Resistance to other antibiotics was documented in some strains of *S. aureus*. In 1996, vancomycin resistance was reported in Japan. In many countries, outbreaks of MRSA infection were reported to be transmitted between hospitals. The rate had increased to 22% by 1995, and by 1997 the level of hospital *S. aureus* infections attributable to MRSA had reached 50%.

The first report of community-associated MRSA (CA-MRSA) occurred in 1981, and in 1982, a large outbreak of CA-MRSA occurred among intravenous drug users in Detroit, Michigan. Additional outbreaks of CA-MRSA were reported through the 1980s and 1990s, including outbreaks among Australian Aboriginal populations that had never been exposed to hospitals. In the mid-1990s, scattered reports of CA-MRSA outbreaks among US children were made. While HA-MRSA rates stabilized between 1998 and 2008, CA-MRSA rates continued to rise. A report released by the University of Chicago Children's Hospital comparing two periods found a 25-fold increase in the rate of hospitalizations due to MRSA among children in the United States. In 1999, the University of Chicago reported the first deaths from invasive MRSA among otherwise healthy children in the United States. By 2004, the genome for various strains of MRSA were described.

The observed increased mortality among MRSA-infected people arguably may be the result of the increased underlying morbidity of these people. Several studies, however, including one by Blot and colleagues, that have adjusted for underlying disease still found MRSA bacteremia to have a higher attributable mortality than methicillin-susceptible *S. aureus* (MSSA) bacteremia.

A population-based study of the incidence of MRSA infections in San Francisco during 2004–05 demonstrated that nearly one in 300 residents suffered from such an infection in the course of a year and that greater than 85% of these infections occurred outside of the healthcare setting. A 2004 study showed that people in the United States with *S. aureus* infection had, on average, three times the length of hospital stay (14.3 vs. 4.5 days), incurred three times the total cost ($48,824 vs. $14,141), and experienced five times the risk of in-hospital death (11.2% vs 2.3%) than people without this infection. In a meta-analysis of 31 studies, Cosgrove *et al.*, concluded that MRSA bacteremia is associated with increased mortality as compared with MSSA bacteremia (odds ratio= 1.93; 95% CI = 1.93 ± 0.39). In addition, Wyllie *et al.* report a death rate of 34% within 30 days among people infected with MRSA, a rate similar to the death rate of 27% seen among MSSA-infected people.

In the US, the CDC issued guidelines on October 19, 2006, citing the need for additional research, but declined to recommend such screening. According to the CDC, the most recent estimates of the incidence of healthcare-associated infections that are attributable to MRSA in the United States indicate a decline in such infection rates. Incidence of MRSA central line-associated blood-stream infections as reported by hundreds of intensive care units decreased 50–70% from 2001–2007. A separate system tracking all hospital MRSA bloodstream infections found an overall 34% decrease between 2005 and 2008. In 2010, vancomycin was the drug of choice.

Across Europe, based mostly on data from 2013, seven countries (Iceland, Norway, Sweden, the Netherlands, Denmark, Finland, and Estonia, from lowest to highest) had low levels of hospital-acquired MRSA infections compared to the others, and among countries with higher levels, significant improvements had been made only in Bulgaria, Poland, and the British Isles.

A 1,000-year-old eye salve recipe found in the medieval *Bald's Leechbook* at the British Library, one of the earliest known medical textbooks, was found to have activity against MRSA *in vitro* and in skin wounds in mice.

MULTIDRUG RESISTANCE IN FUNGI

One of the major advances serving to define the beginning of the era of modern medicine was the development of penicillin in the early 1940s as the first widely used antibiotic effective against microorganisms. In what has become an all-too-common trend, antibiotic-resistant isolates emerged soon after the large-scale use of penicillin to treat bacterial infections. This cycle has been repeated many times over the years as new antibiotics are generated. This has led to obvious concern about the continuing efficacy of antibiotics to control infectious disease.

Resistance to antibiotics in pathogenic fungi is a problem of special importance in the control of infections caused by these organisms. The same extraordinary conservation of the basic eukaryotic cellular biology exhibited by fungal and animal cells that has allowed these smaller eukaryotes to serve as outstanding model organism limits the range of fungus-specific antibiotics that have been described. In addition, mutant fungi are readily isolated, both in the laboratory and in the clinic, that demonstrate resistance to a wide range of antibiotics beyond that initially used for treatment. This broad-spectrum drug tolerance is referred to as multidrug resistance and occurs in organisms ranging from bacteria to humans. The limited number of antifungal drugs makes this phenotype an acute problem in the chemotherapeutic eradication of fungal infections.

Much of our understanding of multidrug resistance in fungi comes from studies in the generally nonpathogenic yeast Saccharomyces cerevisiae, in which the multidrug-resistant phenotype is referred to as pleiotropic drug resistance or Pdr. Genes influencing this phenotype are typically designated PDR loci. With the development of powerful new genetic and molecular biological techniques, workers have provided important new insights into the physiology of multidrug resistance from experiments performed directly in pathogenic organisms.

Pleiotropic Drug Resistance in Saccharomyces Cerevisiae

The facile genetics of S. cerevisiae led to the identification of an allelic series of pleiotropic drug-resistant mutations mapping to a gene present on chromosome VII that define the PDR1 gene. Pdr1p is a zinc cluster-containing positive transcriptional regulator related to the well-known Gal4p transcription factor. Although Pdr1p was the first multidrug resistance determinant identified in S. cerevisiae, this factor is not a direct mediator of drug resistance. Pdr1p and other transcriptional regulators modulate expression of a variety of proteins that act to prevent the toxic action of drugs. We first consider here the proteins that are direct detoxifiers of antifungal agents, then examine the transcription factors regulating these factors, and finally discuss the signals that regulate expression of the PDR genes. We refer here to the set of genes defined by their regulation by Pdr1p and its homologue Pdr3p collectively as the Pdr pathway.

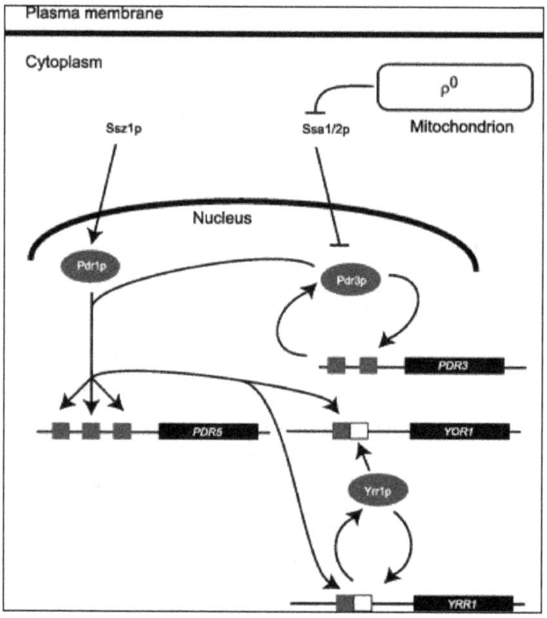

Multidrug resistance pathway in Saccharomyces cerevisiae. A general scheme describing key genetic interactions regulating pleiotropic drug resistance is shown. Standard arrows indicate positive interactions, while the blunt arrows denote negative regulation. The small gray boxes represent Pdr1p/Pdr3p response elements (PDREs), and the white boxes indicate Yrr1p response elements.

Membrane Transporters and other Direct Effectors of Pleiotropic Drug Resistance

The first identified gene that fulfilled the criteria for a direct effector of drug resistance in the Pdr pathway was the PDR5 locus. This gene encodes an ATP-binding cassette transporter protein that is a member of the ABCG class of transporters. Early experiments indicated that the loss of Pdr5p led to a dramatic increase in drug sensitivity to a wide range of different compounds. In addition, direct biochemical assays demonstrated that pdr5Δ cells were defective in the efflux of various dyes and radiolabeled probes. Overproduction of Pdr5p by use of high-copy-number plasmids carrying the gene or the presence of hyperactive PDR1 alleles correlated with a robust Pdr phenotype. Together, these data strongly support the model that elevated Pdr5p levels contribute to the pleiotropic drug-resistant phenotype by increasing the activity of this multispecific drug pump.

Comparison of cells carrying hyperactive PDR1 alleles and containing or lacking PDR5 argued that, although Pdr5p is an important determinant in the Pdr phenotype, the presence of this gene is not sufficient to explain the entire spectrum of drug resistance seen. For example, hyperactive PDR1-6 mutants are resistant to both cycloheximide and oligomycin. Removal of the PDR5 gene eliminates the increased cycloheximide tolerance in a PDR1-6 cell but does not reduce the high-level oligomycin resistance. Screening a high-copy-number plasmid library for genes that influence oligomycin resistance allowed the recovery of the YOR1 gene. YOR1 encodes an ABC transporter of the ABCC family that is required for normal oligomycin resistance. Loss of YOR1 from a PDR1-6 background reduces oligomycin resistance but has no effect on the elevated cycloheximide tolerance conferred by this Pdr1p derivative. Similarly, removal of the PDR5 homologue SNQ2 gene from cells carrying a hyperactive allele of PDR1 reduced 4-nitroquinoline-N-oxide resistance but did not influence cycloheximide or oligomycin resistance.

These data serve to illustrate a major theme in the Pdr phenotype in S. cerevisiae, as well as fungal multidrug resistance in general. Overproduction of multiple ABC transporter proteins is required for the expression of the full range of drug resistance seen in these multidrug-tolerant cells. This typically occurs due to a change in the activity of a transcription factor and shares striking similarity with the multidrug resistance seen in mammalian cells. Isolation of the human MDR1 gene was accomplished by use of cell lines that massively overproduced this multidrug resistance ABC transporter protein. Most multidrug-resistant mammalian cells appear to emerge via amplification of the gene encoding a particular ABC transporter gene for reviews) rather than producing an altered transcription factor but, irrespective of the exact mechanism, both types of cells can become multidrug resistant through the elevation of ABC transporter expression.

Along with the ABC transporter proteins, proteins of the major facilitator superfamily (MFS) also contribute to pleiotropic drug resistance. Compared to the ABC transporter-encoding genes, our understanding of the function and regulation of the MFS proteins is at an earlier stage. At least 20 different MFS proteins exhibit structural characteristics consistent with or have already been demonstrated to play a role in drug resistance. The large number of these membrane transporters suggests that their contributions to drug tolerance are likely underappreciated. The MFS proteins often share overlapping drug specificity with ABC transporters, which may explain in part why MFS transporter involvement in multidrug resistance has been obscured by their better-publicized relatives. For example, the MFS protein Flr1p contributes to resistance to the antifungal agent fluconazole, as well as cycloheximide. The ABC transporter Pdr5p also mediates tolerance to both of these compounds and is clearly a major determinant in the resistance phenotype to these and other drugs. Interestingly, while a pdr5Δ strain is exquisitely sensitive to cycloheximide, this sensitivity can be fully suppressed if the transcription factor Yap1p is overproduced. This suppression is likely due to the activation of FLR1, a Yap1p target gene. Much remains to be learned of the participation of MFS proteins in eukaryotic multidrug resistance.

Sphingolipid Homeostasis is Coregulated with PDR Genes

The lipid composition of the plasma membrane is the central determinant regulating passage of compounds from the external environment to the interior of the cell. The distribution of lipid components in the inner and outer leaflets of the plasma membrane is asymmetric and controlled in part by the Pdr pathway. Both Pdr5p and Yor1p have been found to enhance the outward movement (flop) of the phospholipid phosphatidylethanolamine. Phosphatidylethanolamine is normally maintained at low levels in the outer leaflet by the rapid inward movement (flip) carried out by aminophospholipid translocases, five of which can be found in the S. cerevisiae genome. Intriguingly, two phosphatidylinositol transfer protein homologues (PDR16 and PDR17) were also found to be target genes of the Pdr pathway and to influence phospholipid levels and drug resistance. These observations predict that activation of the Pdr pathway may trigger changes in phospholipid composition of the plasma membrane, but the consequences of these changes remain uncertain.

Eukaryotic membrane lipids include sterols and sphingolipids in addition to phospholipids. In S. cerevisiae, ergosterol is the major sterol in the cell and is produced via the action of ERG pathway. A well-known phenotype of many erg-null mutants is extreme sensitivity to drug challenges, which is consistent with a requirement for ergosterol in normal membrane function. Analyses of erg mutant strains indicate that Pdr5p transport activity is unaffected in these mutants and suggest that loss of ergosterol may enhance passive diffusion of compounds across the altered membrane.

The final class of membrane lipid, sphingolipids, is thought to associate with ergosterol to form microdomains called lipid rafts that are concentrated in the outer leaflet of the plasma membrane. Analysis of the expression of the IPT1 gene, encoding the last step in sphingolipid biosynthesis, established a direct connection between the Pdr pathway and biosynthesis of this class of membrane lipid. IPT1 transcription is controlled by Pdr1p and/or Pdr3p and responds to signals known to induce the S. cerevisiae Pdr pathway. Interestingly, loss of IPT1 altered drug resistance of the resulting mutants and suggests that normal sphingolipid content is required for wild-type levels of drug tolerance. Later experiments provided evidence that several biosynthetic steps upstream of the Ipt1p-catalyzed reaction were also responsive to increases in Pdr pathway-mediated regulation. The known interactions are summarized in figure.

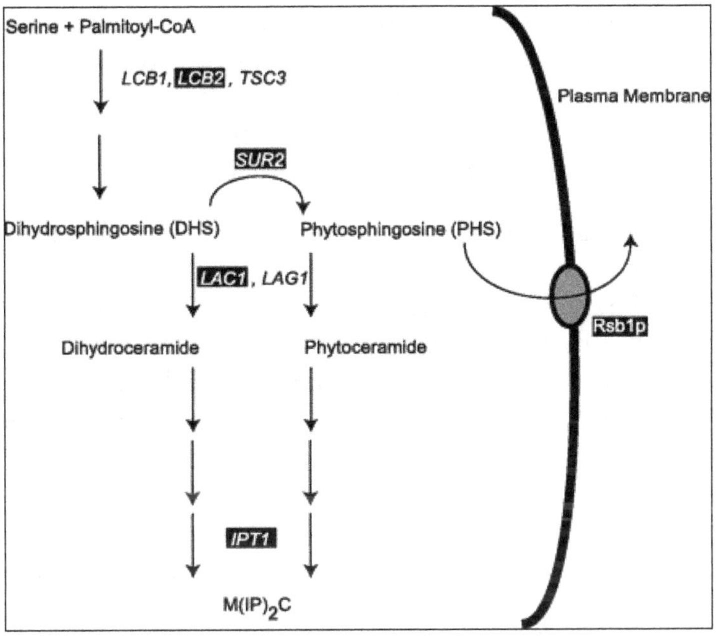

Interactions between the Pdr pathway and sphingolipid biosynthesis. An outline of the biosynthetic pathway producing sphingolipids is depicted. Arrows indicate enzymatic reactions. Genes encoding proteins involved in selected pathway reactions are shown. The black boxes denote genes that are believed to be regulated by Pdr1p and/or Pdr3p. Rsb1p can transport both DHS and PHS. M(IP)2C is mannosyl diinositolphospho-ceramide, the final product of the sphingolipid pathway in S. cerevisiae.

The genetic interactions between the Pdr and sphingolipid pathways were extended by the finding that the RSB1 gene is regulated by Pdr1p and Pdr3p. Rsb1p is thought to act as an effluxer of ceramide precursors called long-chain bases (LCBs). LCBs can be cytotoxic if allowed to accumulate, and Rsb1p can act to prevent inappropriate buildup of these sphingolipid intermediates. Surprisingly, loss of Pdr5p and Yor1p from cells strongly elevates LCB resistance in a Rsb1p-dependent fashion, again emphasizing the interconnections between sphingolipid biosynthesis and the Pdr pathway. These multiple interfaces between pleiotropic drug resistance and the homeostasis of membrane lipids suggest the possibility that the physiological role of the Pdr pathway is to assist in regulation of the function of the plasma membrane at the level of both lipids and membrane transporters.

Transcriptional Control of Pleiotropic Drug Resistance

A central determinant in the drug resistance phenotype of S. cerevisiae is provided by the regulation of transactivation capability of a limited number of regulatory proteins. Pdr1p and its homologue Pdr3p are Zn_2Cys_6-containing transcriptional regulatory proteins that exert major influences on the multidrug resistance phenotypes of cells. There are a large number of Zn_2Cys_6-containing transcription factors in S. cerevisiae (>50), and many of these have been shown or are believed to be involved in pleiotropic drug resistance. We will focus on Pdr1p and Pdr3p as illustrative of the larger number of Zn_2Cys_6-containing factors that contribute to drug resistance. More detailed considerations of these and other transcriptional regulatory proteins involved in multidrug resistance in S. cerevisiae are available in several reviews.

Hyperactive mutant forms of Pdr1p drove the initial identification of the pleiotropic drug resistance phenotype in S. cerevisiae. Similar mutant alleles of PDR3 have also been described. These single amino acid substitution forms of Pdr1p and Pdr3p produce transcriptional regulatory proteins that behave as strong, constitutive activators of downstream gene expression. Pdr1p and Pdr3p both bind to elements referred to as Pdr1p/Pdr3p response elements (PDREs) located upstream of target genes. In vivo footprinting experiments indicate that Pdr1p and Pdr3p are likely to be constitutively bound to relevant PDREs, an observation consistent with the constitutive nuclear localization of these proteins. In addition, both Pdr1p and Pdr3p can activate the expression of another zinc cluster transcription factor-encoding gene called YRR1. Increased expression of Yrr1p can amplify the transcriptional effects of activation of either Pdr1p or Pdr3p since this factor recognizes a sequence different from the PDRE.

Although significant similarities are shared by Pdr1p and Pdr3p, important differences are known. First, these factors are expressed at dramatically different levels. Use of epitope-tagged forms of both proteins indicates that Pdr1p is present at nearly 10 times the level of Pdr3p. Second, the regulation of these factors is responsive to different signals. Overproduction of the DnaK protein Ssz1p or the DnaJ Zuo1p induces Pdr1p-dependent gene transcription but has no effect on Pdr3p. Conversely, cells lacking their mitochondrial genome (ρ°) activate Pdr3p function but have no effect on Pdr1p. PDR1 expression levels are constant, whereas PDR3 is both autoregulated and highly induced in ρ° cells. Finally, recent work from our laboratory demonstrates the Hsp70 protein Ssa1p can negatively regulate Pdr3p but not Pdr1p activity. Although these two transcription factors share >30% identity across their roughly 1,000-amino-acid lengths, these differences indicate that Pdr1p and Pdr3p have nonidentical roles to play in the control of multidrug resistance.

The second class of transcription factor that has been associated with the regulation of multidrug resistance is the basic region-leucine zipper (bZip) family of regulatory proteins. Although a number of these bZip-containing factors are present in S. cerevisiae, we will restrict to Yap1p, the first of these factors shown to be involved in pleiotropic drug resistance. Yap1p is better known for its important role in oxidative stress tolerance, and its regulation by oxidants has been the subject of intensive study. Briefly, Yap1p normally cycles between the nucleus and the cytoplasm in the absence of stress. Upon oxidant challenge, Yap1p rapidly accumulates in the nucleus, where it activates antioxidant gene expression. Mutant forms of Yap1p have been described that are constitutively located in the nucleus and hyper-resistant to certain oxidants.

Less is known of the response mediated by Yap1p upon drug challenge, but the YAP1 gene was isolated as a high-copy-number mediator of pleiotropic drug resistance along with PDR5. As mentioned above, Yap1p defines a pathway for drug resistance parallel to that of PDR5. Interestingly, Yap1p is known to activate the expression of at least two different MFS protein-encoding genes: ATR1 and FLR1. At least in S. cerevisiae, the control of multidrug resistance gene expression appears to be divided between zinc cluster-containing factors primarily driving transcription of ABC transporter-encoding genes, while bZip-containing proteins mainly act by regulating mRNA levels of MFS proteins. The rationale underlying this division of transcriptional regulatory circuits remains to be determined.

Candida Albicans

While the powerful genetics associated with the use of S. cerevisiae as a model fungus have allowed the most rapid progress in our understanding of fungal multidrug resistance, this organism is not a significant cause of human disease. The major fungal source of bloodstream infections in humans is from the genus Candida which has grown in importance until candidemia now represents the fourth most common nosocomial infection. The primary Candida species associated with candidemia is Candida albicans, and this organism has been the most intensively studied in terms of multidrug resistance.

The first multidrug resistance gene from C. albicans was recovered by selection of a segment of C. albicans genomic DNA that altered drug resistance in S. cerevisiae when carried on an appropriate plasmid. The first gene isolated in this fashion was designated BENr and is now referred to as MDR1. The protein encoded by this gene is a member of the major facilitator superfamily of membrane transporters. Shortly after this finding, two genes encoding homologues of the S. cerevisiae Pdr5p ABC transporter protein were identified. These proteins were given the acronyms CDR1 and CDR2 for Candida drug resistance 1 and 2. Experiments in several labs established that mutant C. albicans strains lacking these ABC transporters were multidrug sensitive and that a number of different resistant isolates were found to overproduce CDR1 and CDR2 transcripts. Measurements of drug transport in strains engineered to overproduce Cdr1p support the idea that this protein acts as an ATP-dependent drug efflux pump. Green fluorescent protein fusions to Cdr1p indicated that this protein was primarily found in the plasma membrane, similar to the location of Pdr5p in S. cerevisiae.

More recent work has focused on the control of transcription of CDR1 and CDR2. Experiments from two different laboratories provided characterization of the promoter region of CDR1. CDR1 transcriptional control was found to be complex and the product of multiple DNA elements of either positive or negative nature in the promoter region. Analysis of the CDR2 promoter suggests that this gene is regulated in parallel with CDR1 since azole-resistant C. albicans isolates often exhibit elevated mRNA levels corresponding to both of these ABC transporter-encoding genes.

Several different transcriptional regulatory proteins have been implicated in the modulation of CDR1 and CDR2 expression. The best characterized of these is the Zn_2Cys_6 cluster protein Tac1p. This factor binds to a single element in the CDR2 promoter that contains tandem repeats of a CGG sequence. These short trinucleotide repeats are commonly associated with the binding sites of Zn_2Cys_6 cluster proteins. Tac1p exhibits the highest degree of sequence similarity with a protein from S. cerevisiae designated Hal9p, a factor involved in the control of expression of the ENA1 sodium-potassium ATPase. Interestingly, the presence of a CDR2-lacZ fusion in S. cerevisiae does not lead to the production of significant β-galactosidase activity unless Tac1p is heterologously provided, suggesting that Hal9p cannot stimulate CDR2 expression. Although Pdr1p does share significant sequence similarity with Tac1p (E value = 10^{-13}), there are clear differences between these two Zn_2Cys_6 cluster proteins at the level of DNA binding specificity and protein sequence. An interesting commonality between PDR1 and TAC1 gain-of-function mutants is the spectrum of highly responsive target genes that these two important regulators of multidrug resistance induce in their respective organisms. In S. cerevisiae, PDR1 hyperactive mutants strongly induce PDR5 and PDR15, as well as another membrane protein called Rsb1p that is required for resistance to the long-chain base phytosphingosine. In C. albicans, TAC1 hyperactive alleles induce CDR1 (Pdr5p homologue), CDR2 (Pdr15p homologue), and RTA3 (Rsb1p homologue).

A second positive regulator of CDR1 was found by screening a C. albicans expression library in a S. cerevisiae cell carrying a CDR1-lacZ reporter gene for clones that could elevate expression of this heterologous reporter gene. This gene was designated CaNDT80 since it was found to encode a homologue of the S. cerevisiae Ndt80p transcription factor. ScNdt80p serves to activate genes involved in sporulation in S. cerevisiae. CaNdt80p appears to have functionally diverged from the role of its S. cerevisiae cognate protein since the C. albicans factor is important in positive regulation of CDR1 during vegetative growth.

Evidence implicating an important role for a negative regulator of CDR1 expression has also accrued. To date, the identity of this factor is still undetermined, but its binding site has been mapped to an element located downstream of the Tac1p recognition element. A 55-kDa protein has been shown to be cross-linked to this negative regulatory element, and reductions in the level of this factor are believed to cause azole hyper-resistance in some clinical isolates. Together, these data suggest that control of expression of CDR1 in C. albicans involves more complexity at the level of the transcription factors compared to the control of ScPDR5 expression that is primarily dependent on only Pdr1p and Pdr3p.

Transcriptional control of CaMDR1 has also been examined by several laboratories. Detailed deletion mapping experiments have been described in several publications. These studies have identified several different regions in the CaMDR1 promoter that are involved in basal, oxidant- or drug-induced expression. There are differences in the precise roles of these multiple elements but clearly the CaMDR1 promoter is a complex transcriptional control region that integrates multiple inputs to determine the proper expression of CaMdr1p.

At least two different factors have been linked to the trans-regulation of CaMDR1 expression. Isolation of the C. albicans Yap1p homologue (Cap1p) indicated that CaMDR1 was a likely downstream target of this transcription factor. This has been supported by direct mutagenesis of putative Cap1p regulatory elements present in CaMDR1. Surprisingly, other analyses demonstrated that loss of CAP1 either had no effect on azole resistance or negatively influenced tolerance to this drug. However, overproduction of a carboxy-terminal truncation mutant of Cap1p led to a dramatic increase in azole resistance, an observation that correlates with strongly elevated CaMDR1 transcription. This behavior is similar to that observed in S. cerevisiae when YAP1 is deleted as yap1Δ cells exhibit a modest (twofold) increase while overproduction of Yap1p produced a striking increase in the MIC for fluconazole.

Along with Cap1p, the C. albicans homologue of S. cerevisiae Mcm1p has also been implicated in control of CaMDR1 transcription. Laboratory strains of C. albicans that were selected for high azole resistance were found to be dependent on the presence of a functional CaMcm1p binding site for high level production of CaMDR1. Cap1p was not required for this effect arguing that CaMcm1p is the primary if not the sole factor that is upregulated in these azole-resistant isolates. Azole resistance has also been shown to be impacted by changes in expression of the C. albicans IPT1 gene, suggesting a further conservation with S. cerevisiae. The picture that emerges from studies in C. albicans on the molecular basis of multidrug resistance is consistent with this organism sharing a very similar range of effector genes with S. cerevisiae, but with the regulation of gene expression exhibiting differences. This may be due to the very different milieus in which these organisms normally reside. Further study of the resistance pathways in these yeasts will clarify this notion.

Candida Glabrata

C. glabrata is a haploid species of Candida that has emerged as the second most common Candida organism associated with fungemia. A likely contributing factor to the rapid growth (2% in the 1970s to 20% now) is the robust ability of C. glabrata to acquire tolerance to commonly deployed antifungal agents such as azoles. Long-term monitoring of C. glabrata-associated disease indicates that azole-resistant isolates are increasing in frequency, even from geographic regions in which C. glabrata was originally sensitive to these drugs. This is not seen in the case of C. albicans since clinical data indicate that azole drugs retain their efficacy in controlling disease associated with this species. A second complicating factor for C. glabrata comes from the routine multidrug resistance of azole-tolerant isolates. Not only does C. glabrata relatively easily convert to an azole-resistant form, but it also frequently appears to become simultaneously multidrug resistant.

Protein and DNA sequence similarity analyses indicate that C. glabrata and S. cerevisiae are closely related organisms. This close relationship is also illustrated by the similarities in the multidrug resistance pathways in these two yeasts. ρ° C. glabrata cells are highly multidrug resistant. This mitochondrial control of multidrug resistance proceeds through the activation of the CgPdr1p protein. A notable difference between S. cerevisiae and C. glabrata is that CgPdr1p represents the single best homologue of the Pdr1p/Pdr3p pair of proteins found in S. cerevisiae. CgPDR1 may correspond to a fusion gene between the S. cerevisiae PDR1 and PDR3 loci. CgPDR1 shares more sequence similarity with ScPdr1p but exhibits the ρ° induction seen for ScPDR3. Single amino acid substitution mutant forms of CgPdr1p have been found that are linked with high-level transcription of both CgPDR1 and CgCDR1, as well as robust multidrug resistance. This behavior is directly analogous to that seen for the hyperactive alleles of ScPDR1 and ScPDR3.

Although it is clear that ρ° C. glabrata cells are highly multidrug resistant and occur frequently, the role of these cells in disease is less certain. C. glabrata clinical isolates that were converted to petite status by either ethidium bromide or fluconazole treatment were found to have reduced virulence in a mouse model of fungal pathogenesis. One concern with these experiments is that evaluation of virulence was carried out without the administration of fluconazole. Since the absence of drug selection would remove the possible selective advantage provided by the high-level expression of the CgCDR genes and associated multidrug resistance, the petite cells may be lost due to competition from the healthy endogenous microbial flora. There have been reports of petite C. glabrata strains isolated from patients, but a larger number of other clinical isolates are not apparently linked with a petite character.

While the finding that petite strains of C. glabrata and S. cerevisiae exhibit a multidrug-resistant phenotype illustrates the similar mitochondrial regulatory basis in these organisms, an interesting difference has been uncovered by screening a collection of transposon-generated C. glabrata mutants. Insertion of a transposon into several genes was found to elevate azole resistance but also trigger petite formation. When the selective pressure was removed, the petite phenotype appeared to abate. Previous studies on both C. glabrata and S. cerevisiae petite mutants have not uncovered multidrug-resistant mutants that exhibit reversible behavior.

One feature of the C. glabrata petite multidrug-resistant mutants that remains unclear is the status of their mitochondrial genome. The so-called high-frequency azole-resistant mutants were demonstrated to lack a mitochondrial genome and closely resemble the ρ° multidrug-resistant mutants of S. cerevisiae. Three of the transposon-induced C. glabrata petite mutants were examined for mitochondrial genome status and intriguingly found to still retain cytoplasmic nucleoids. This is an important potential difference since a number of different S. cerevisiae mutants that are petite but not ρ° fail to express the multidrug-resistant phenotype. The petite transposon insertion mutants in C. glabrata include an insertion in the CgSHE9 gene. Disruption mutants lacking SHE9 (aka MDM33) do not appear to lose its mitochondrial genome in either S. cerevisiae or C. glabrata, but multidrug resistance has only been assessed for C. glabrata mutants. Further work is required to determine whether the reversible multidrug-resistant status of the three C. glabrata genes is also exhibited by their cognate S. cerevisiae mutants. As suggested previously, mutants that exhibit a gradual diminution of mitochondrial function without complete loss of the mitochondrial genome may give rise to this reversible multidrug resistance. Mutants of this type would have clear advantages in terms of survival of chemotherapy in a patient since they could combine the benefits of robust multidrug resistance of a petite cell without permanently acquiring the fragile growth of these mitochondrially deficient strains.

Multidrug resistance in C. glabrata is also influenced by transcriptional control of MFS protein expression. The CgFLR1 gene is induced by the C. glabrata Yap1p homologue (CgAP-1) and confers resistance to a range of agents, including fluconazole. C. glabrata Cgap-1 mutants had normal fluconazole tolerance, while Cgpdr1 mutants were hypersensitive to this drug. The Yap1p-FLR1 fluconazole resistance pathway seems to serve an ancillary role in tolerance to this antifungal agent in both S. cerevisiae and C. glabrata. Somewhat surprisingly, while Cgap-1 mutants are hypersensitive to oxidants, there is no detectable effect on virulence. Although CgAP-1 is an important determinant of oxidative stress resistance, this does not appear to influence the ability of C. glabrata to colonize a mouse model.

The similarities between multidrug-resistant isolates of S. cerevisiae and C. glabrata have been extended by analysis of a genes that are transcriptionally responsive to a hyperactive mutant form of CgPdr1p by microarray. While a number of genes that are induced in C. glabrata represent homologues that are also elevated in response to S. cerevisiae hyperactive Pdr1p, the largest group of induced transcripts correspond to genes that are uniquely induced in the pathogenic yeast. These C. glabrata-specific genes may represent loci that are required for the organism to successfully withstand the challenges of toxic compounds when exposed in an animal host.

Aspergillus Spp.

The primary human pathogen among the filamentous fungi is Aspergillus fumigatus. Infections associated with this organism are have a high morbidity and an intrinsically high resistance to the standard cadre of antifungal agents. Not surprisingly, continued use of azole-based drugs has the undesirable consequence of elevating the resistance of subsequent isolates from these patients. Fortunately, acquisition of a multidrug resistance phenotype is relatively rare among Aspergillus species, but this statement must be qualified by the recognition that analysis of this phenotype is complicated by technical difficulties in establishing this phenotype. In addition, molecular characterization of genes involved in Aspergillus multidrug resistance has been hindered by lack of a complete genomic sequence. This has recently been accomplished and will accelerate study of the loci involved in drug resistance in this important human pathogen.

Previous studies have provided some information concerning multidrug resistance genes in Aspergillus. Degenerate PCR cloning allowed the isolation of two different ABC transporter-encoding genes, and expression of one of these in S. cerevisiae elevated resistance to an antifungal drug of the echinocandin family. Using a probe for C. albicans CDR1, A. fumigatus atrF was isolated and shown to be inducible in azole-resistant isolates but not inducible in azole-susceptible strains. Azole-resistant isolates selected in vitro were found to have elevated expression of ABC transporter and MFS-encoding genes. These data are consistent with similar systems for drug detoxification existing in Aspergillus as have been described in yeasts.

The availability of the genomic sequence of Aspergillus has already provided interesting insights into the likely composition of multidrug resistance systems in this filamentous fungus. The A. fumigatus genome is ca. 50% larger than that of S. cerevisiae and yet predicts the presence of 96 potential MFS multidrug transporters compared to 24 from budding yeast. Similarly, S. cerevisiae contains 13 ABC transporter genes thought to participate in multidrug resistance, whereas A. fumigatus is believed to have 35. Analyses of these proteins in A. fumigatus is required to confirm that these proteins are involved in drug resistance in this organism, but the presence of such a proportionally larger number suggests a hardening of the drug defensive capability of this pathogenic organism. It is striking to note that the increased proportion of MFS proteins compared to ABC transporters found in the Aspergillus genome compared to S. cerevisiae resembles the situation in pathogenic bacteria. Currently, there are no examples of ABC transporters in clinically relevant bacteria pathogens that are important in drug resistance, whereas MFS protein participation in evasion of antibacterial therapy is very common.

References

- Graham M., Snyder; O'Fallon, E; D'Agata, EMC (21 Mar 2010). Risk Factors for Co-Colonization with Multidrug-Resistant Gram-Negative Bacteria in a Long-Term Care Facility. Atlanta: International Conference on Healthcare-Associated Infections. Retrieved 1 Dec, 2010
- Lewis K (2007). "Persister cells, dormancy and infectious disease". Nature Reviews Microbiology. 5 (1): 48–56. Doi:10.1038/nrmicro1557. PMID 17143318
- Robinson, D. Ashley; Feil, Edward J.; Falush, Daniel (2010-03-16). Bacterial Population Genetics in Infectious Disease. John Wiley & Sons. ISBN 9780470600115
- Chen, Luke F. (23 Nov 2010). "The Increasing Threat of Multidrug-Resistant Gram-Negative In-fections". Medscape Medical News. Webmd. Retrieved 1 Dec, 2010
- Winn, Washington (2006). Koneman's color atlas and textbook of diagnostic microbiology. Philadelphia: Lippincott Williams & Wilkins. ISBN 978-0781730143
- Brauner, A (2017). "Distinguishing between resistance, tolerance and persistence to antibiotic treatment". Nature Reviews Microbiology. 14 (5): 320–30. Doi:10.1038/nrmicro.2016.34. PMID 27080241
- Gould, Dinah; Brooker, Christine (2008-08-20). Infection Prevention and Control: Applied Microbiology for Healthcare. Palgrave Macmillan. ISBN 9781137045928

5

Diverse Aspects of Antimicrobial Drug Resistance

There are many important concepts that are studied under the domain of antimicrobial drug resistance such as horizontal gene transfer, eagle effect, selectivity factor and r-factor. This chapter has been carefully written to provide an easy understanding of these related concepts of antimicrobial drug resistance.

HORIZONTAL GENE TRANSFER

Bacteria are able to respond to selective pressures and adapt to new environments by acquiring new genetic traits as a result of mutation, a modification of gene function within a bacterium, and as a result of horizontal gene transfer, the acquisition of new genes from other bacteria. Mutation occurs relatively slowly. The normal mutation rate in nature is in the range of 10-6 to 10-9 per nucleotide per bacterial generation, although when bacterial populations are under stress, they can greatly increase their mutation rate. Furthermore, most mutations are harmful to the bacterium. Horizontal gene transfer, on the other hand, enables bacteria to respond and adapt to their environment much more rapidly by acquiring large DNA sequences from another bacterium in a single transfer.

Horizontal gene transfer, also known as lateral gene transfer, is a process in which an organism transfers genetic material to another organism that is not its offspring. The ability of Bacteria and Archaea to adapt to new environments as a part of bacterial evolution most frequently results from the acquisition of new genes through horizontal gene transfer rather than by the alteration of gene functions through mutations. (It is estimated that as much as 20% of the genome of Escherichia coli originated from horizontal gene transfer.)

Horizontal gene transfer is able to cause rather large-scale changes in a bacterial genome. For example, certain bacteria contain multiple virulence genes called pathogenicity islands that are located on large, unstable regions of the bacterial genome. These pathogenicity islands can be transmitted to other bacteria by horizontal gene transfer. However, if these transferred genes provide no selective advantage to the bacteria that acquire them, they are usually lost by deletion. In this way the size of the bacterium's genome can remain approximately the same size over time.

There are three mechanisms of horizontal gene transfer in bacteria: transformation, transduction, and conjugation. The most common mechanism for horizontal gene transmission among bacteria, especially from a donor bacterial species to different recipient species, is conjugation. Although bacteria can acquire new genes through transformation and transduction, this is usually a more rare transfer among bacteria of the same species or closely related species.

Transformation

Transformation is a form of genetic recombination in which a DNA fragment from a dead, degraded bacterium enters a competent recipient bacterium and is exchanged for a piece of DNA of the recipient. Transformation usually involves only homologous recombination, a recombination of homologous DNA regions having nearly the same nucleotide sequences. Typically this involves similar bacterial strains or strains of the same bacterial species.

A few bacteria, such as Neisseria gonorrhoeae, Neisseria meningitidis, Hemophilus influenzae, Legionella pneomophila, Streptococcus pneumoniae, and Helicobacter pylori tend to be naturally competent and transformable. Competent bacteria are able to bind much more DNA than noncompetent bacteria. Some of these genera also undergo autolysis that then provides DNA for homologous recombination. In addition, some competent bacteria kill noncompetent cells to release DNA for transformation.

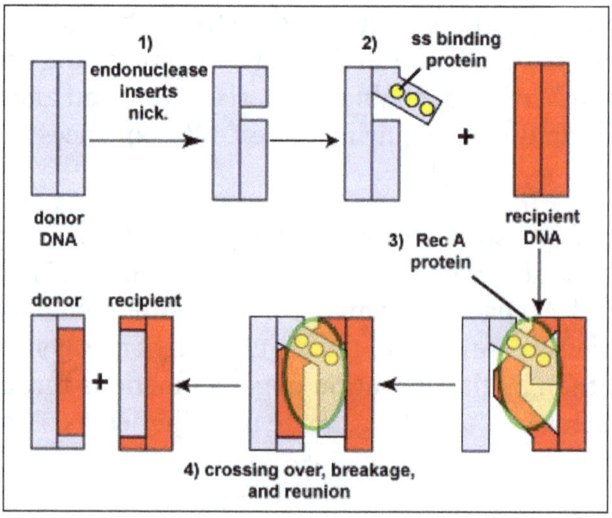

Pairing of Homologous DNA molecules and Exchange of DNA Segments by way of Rec A Protein. 1) A DNA endonuclease inserts a nick in one strand of the donor DNA. 2) The nicked strand is separated from its partner strand by proteins functioning as a helicase. Molecules of single-stranded binding protein (yellow) then bind. 3) Rec A protein then binds to the single-strand fragment and promotes base pairing of the donor DNA with the recipient DNA (crossing over). 4) The linked molecules are separated by resolvases, enzymes that cut and rejoin the cross-linked DNA molecules.

During transformation, DNA fragments (usually about 10 genes long) are released from a dead degraded bacterium and bind to DNA binding proteins on the surface of a competent living recipient bacterium. Depending on the bacterium, either both strands of DNA penetrate the recipient, or a nuclease degrades one strand of the fragment and the remaining DNA strand enters the recipient. This DNA fragment from the donor is then exchanged for a piece of the recipient's DNA by means of RecA proteins and other molecules and involves breakage and reunion of the paired DNA segments as seen in.

Step 1: A donor bacterium dies and is degraded.

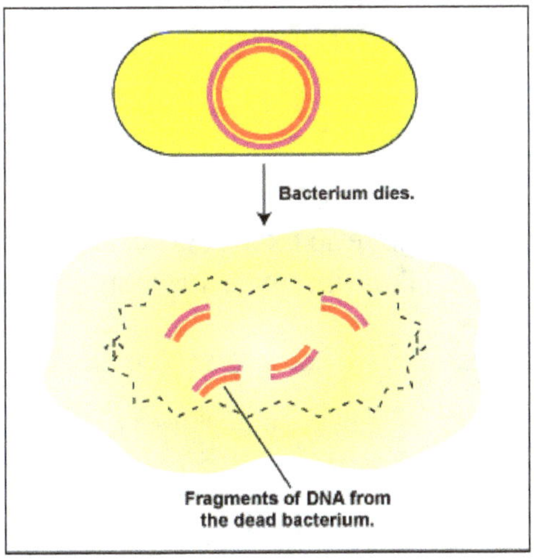

Step 2: DNA fragments, typically around 10 genes long, from the dead donor bacterium bind to transformasomes on the cell wall of a competent, living recipient bacterium.

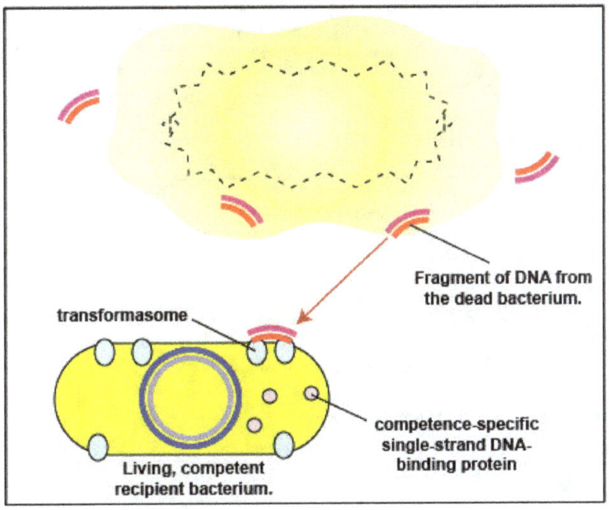

Step 3: In this example, a nuclease degrades one strand of the donor fragment and the remaining DNA strand enters the recipient. Competence-specific single-stranded DNA-binding proteins bind to the donor DNA strand to prevent it from being degraded in the cytoplasm.

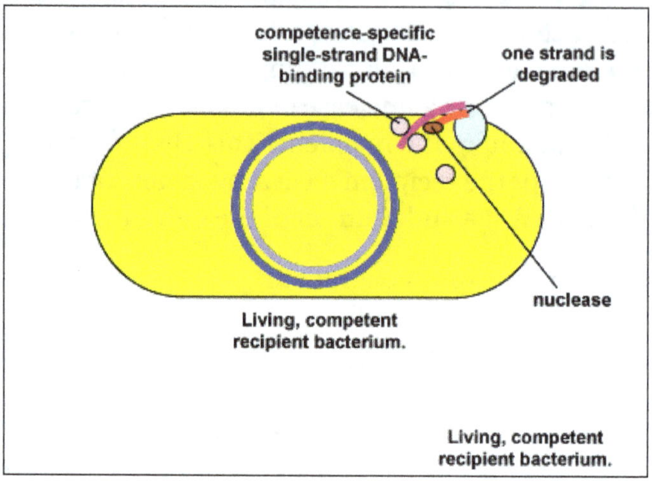

Step 4: RecA proteins promotes genetic exchange between a fragment of the donor's DNA and the recipient's DNA. This involves breakage and reunion of paired DNA segments.

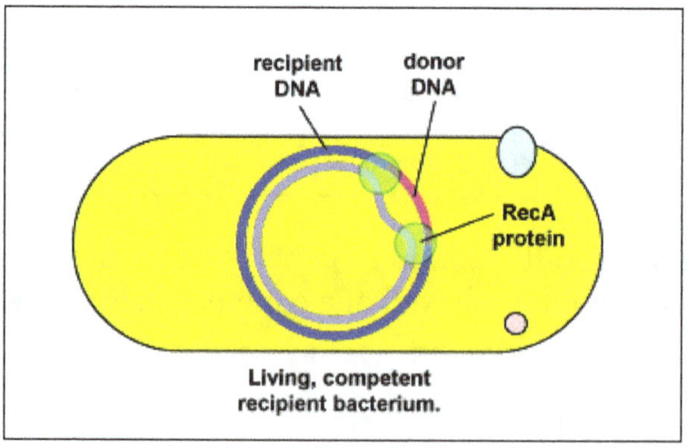

Step 5: Transformation is complete.

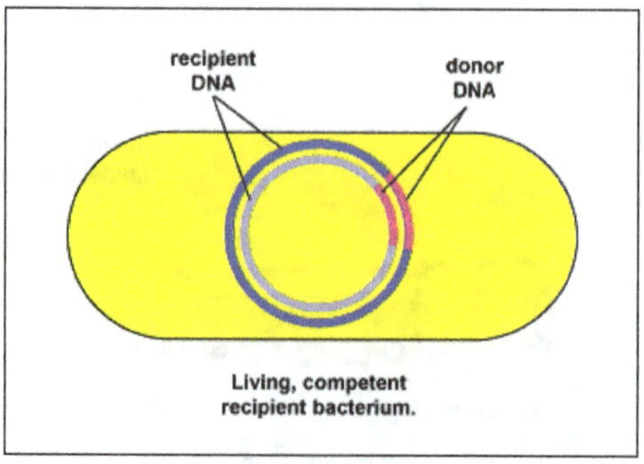

Transduction

Transduction involves the transfer of a DNA fragment from one bacterium to another by a bacteriophage. There are two forms of transduction: generalized transduction and specialized transduction.

During the replication of lytic bacteriophages and temperate bacteriophages, occasionally the phage capsid accidently assembles around a small fragment of bacterial DNA. When this bacteriophage, called a transducing particle, infects another bacterium, it injects the fragment of donor bacterial DNA it is carrying into the recipient where it can subsequently be exchanged for a piece of the recipient's DNA by homologous recombination.

- Step 1: A bacteriophage adsorbs to a susceptible bacterium.
- Step 2: The bacteriophage genome enters the bacterium. The genome directs the bacterium's metabolic machinery to manufacture bacteriophage components and enzymes. Bacteriophage-coded enzymes will also breakup the bacterial chromosome.
- Step 3: Occasionally, a bacteriophage capsid mistakenly assembles around either a fragment of the donor bacterium's chromosome or around a plasmid instead of around a phage genome.
- Step 4: The bacteriophages are released as the bacterium is lysed. Note that one bacteriophage is carrying a fragment of the donor bacterium's DNA rather than a bacteriophage genome.
- Step 5: The bacteriophage carrying the donor bacterium's DNA adsorbs to a recipient bacterium.
- Step 6: The bacteriophage inserts the donor bacterium's DNA it is carrying into the recipient bacterium.
- Step 7: Homologous recombination occurs and the donor bacterium's DNA is exchanged for some of the recipient's DNA.

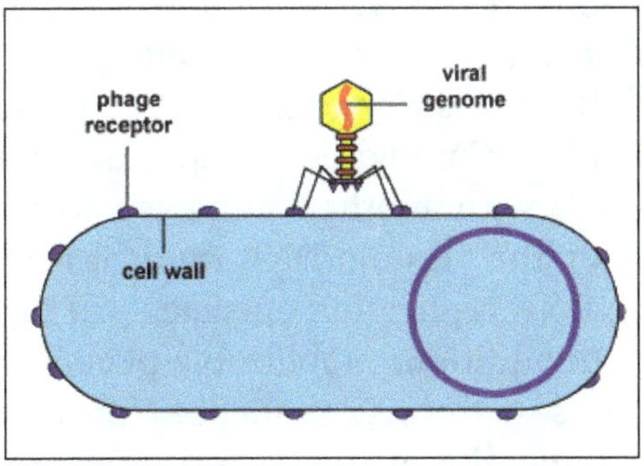

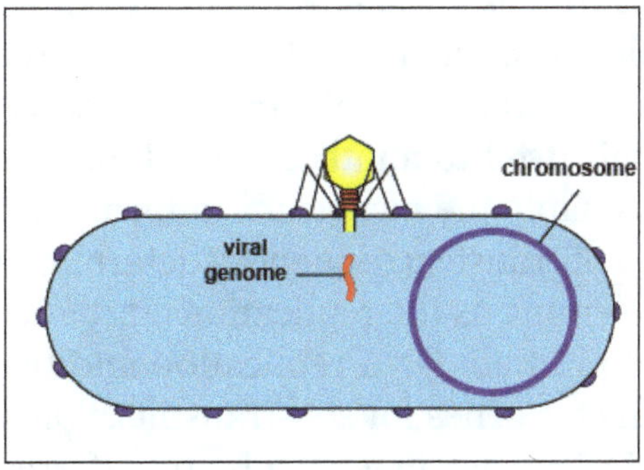

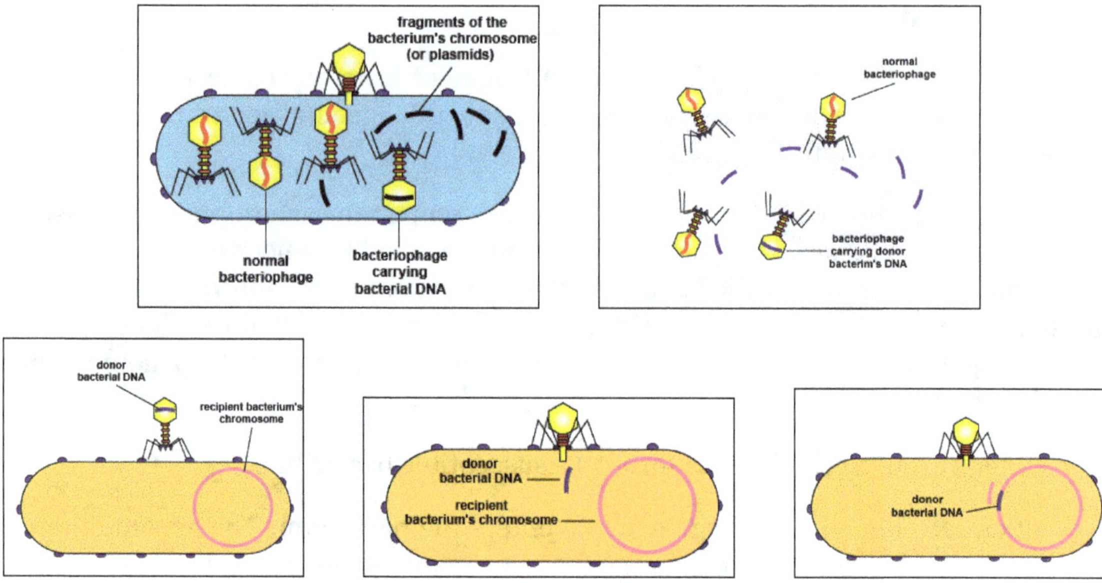

Generalized transduction by lytic bacteriophage.

Plasmids, such as the penicillinase plasmid of Staphylococcus aureus, may also be carried from one bacterium to another by generalized transduction.

Specialized transduction: This may occur occasionally during the lysogenic life cycle of a temperate bacteriophage. During spontaneous induction, a small piece of bacterial DNA may sometimes be exchanged for a piece of the bacteriophage genome, which remains in the bacterial nucleoid. This piece of bacterial DNA replicates as a part of the bacteriophage genome and is put into each phage capsid. The bacteriophages are released, adsorb to recipient bacteria, and inject the donor bacterium DNA/phage DNA complex into the recipient bacterium where it inserts into the bacterial chromosome.

Conjugation

Genetic recombination in which there is a transfer of DNA from a living donor bacterium to a living recipient bacterium by cell-to-cell contact. In Gram-negative bacteria it typically involves a conjugation or sex pilus.

Conjugation is encoded by plasmids or transposons. It involves a donor bacterium that contains a conjugative plasmid and a recipient cell that does not. A conjugative plasmid is self-transmissible, in that it possesses all the necessary genes for that plasmid to transmit itself to another bacterium by conjugation. Conjugation genes known as tra genes enable the bacterium to form a mating pair with another organism, while oriT (origin of transfer) sequences determine where on the plasmid DNA transfer is initiated by serving as the replication start site where DNA replication enzymes will nick the DNA to initiate DNA replication and transfer. In addition, mobilizable plasmids that lack the tra genes for self-transmissibility but possess the oriT sequences for initiation of DNA transfer may also be transferred by conjugation if the bacterium containing them also possesses a conjugative plasmid. The tra genes of the conjugative plasmid enable a mating pair to form, while the oriT of the mobilizable plasmid enable the DNA to moves through the conjugative bridge.

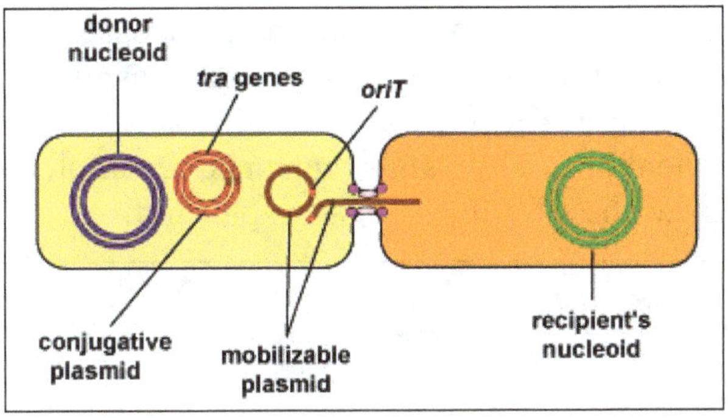

Transfer of mobilizable plasmids during conjugation. Mobilizable plasmids, that lack the tra genes for self-transmissibility but possess the oriT sequences for initiation of DNA transfer, may also be transferred by conjugation if the bacterium containing them also possesses a conjugative plasmid. The tra genes of the conjugative plasmid enable a mating pair to form while the oriT quences of the mobilizable plasmid enables the DNA to move through the conjugative bridge.

Transposons ("jumping genes") are small pieces of DNA that encode enzymes that enable the transposon to move from one DNA location to another, either on the same molecule of DNA or on a different molecule. Transposons may be found as part of a bacterium's chromosome (conjugative transposons) or in plasmids and are usually between one and twelve genes long. A transposon contains a number of genes, such as those coding for antibiotic resistance or other traits, flanked at both ends by insertion sequences coding for an enzyme called transpoase. Transpoase is the enzyme that catalyzes the cutting and resealing of the DNA during transposition.

Conjugative transposons, like conjugative plasmids, carry the genes that enable mating pairs to form for conjugation. Therefore, conjugative transposons also enable mobilizable plasmids and nonconjugative transposons to be transferred to a recipient bacterium during conjugation.

Many conjugative plasmids and conjugative transposons possess rather promiscuous transfer systems that enables them to transfer DNA not only to like species, but also to unrelated species. The ability of bacteria to adapt to new environments as a part of bacterial evolution most frequently results from the acquisition of large DNA sequences from another bacterium by conjugation.

General mechanism of transfer of conjugative plasmids by conjugation in Gram-negative bacteria: In Gram-negative bacteria, the first step in conjugation involves a conjugation pilus (sex pilus or F pilus) on the donor bacterium binding to a recipient bacterium lacking a conjugation pilus. Typically the conjugation pilus retracts or depolymerizes pulling the two bacteria together. A series of membrane proteins coded for by the conjugative plasmid then forms a bridge and an opening between the two bacteria, now called a mating pair.

Using the rolling circle model of DNA replication, a nuclease breaks one strand of the plasmid DNA at the origin of transfer site (oriT) of the plasmid and that nicked strand enters the recipient bacterium. The other strand remains behind in the donor cell. Both the donor and the recipient plasmid strands then make a complementary copy of themselves. Both bacteria now possess the conjugative plasmid.

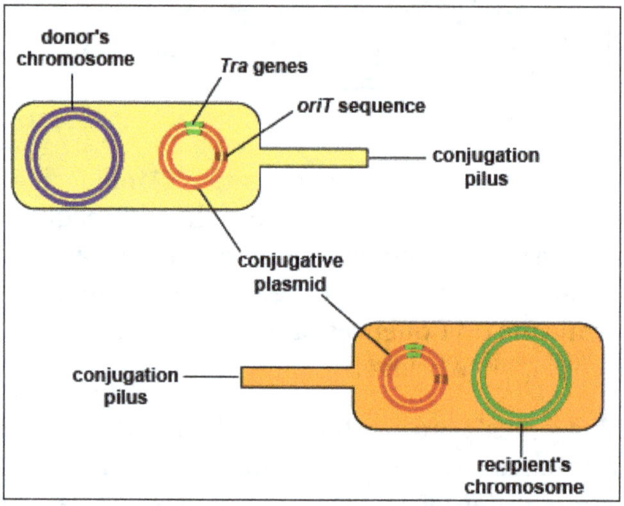

Transfer of conjugative plasmids. Step 1: In Gram-negative bacteria, the first step in conjugation involves a conjugation pilus (sex pilus or F pilus) on the donor bacterium binding to a recipient bacterium lacking a conjugation pilus. Step 2: Typically the conjugation pilus retracts or depolymerizes pulling the two bacteria together. A series of membrane proteins coded for by the conjugative plasmid then forms a bridge and an opening between the two bacteria, now called a mating pair. Step 3: Using the rolling circle model of DNA replication, a nuclease breaks one strand of the plasmid DNA at the origin of transfer site (oriT) of the plasmid. The nuclease also has helicase activity and unwinds the strand that is going to be transferred. Step 4: The nicked plasmid strand enters the recipient bacterium. The other strand remains behind in the donor cell. Step 5: Both the donor and the recipient plasmid strands then make a complementary copy of themselves. Step 6: Both bacteria now possess the conjugative plasmid and can make a conjugation pilus.

This is the mechanism by which resistance plasmids (R-plasmids), coding for multiple antibiotic resistance and conjugation pilus formation, are transferred from a donor bacterium to a recipient. This is a big problem in treating opportunistic Gram-negative infections such as urinary tract infections, wound infections, pneumonia, and septicemia by such organisms as E. coli, Proteus, Klebsiella, Enterobacter, Serratia, and Pseudomonas, as well as with intestinal infections by organisms like Salmonella and Shigella.

There is also evidence that the conjugation pilus may also serve as a direct channel through which single-stranded DNA may be transferred during conjugation.

F^+ Conjugation

This results in the transfer of an F^+ plasmid possessing tra genes coding only for a conjugation pilus and mating pair formation from a donor bacterium to a recipient bacterium. One strand of the F^+ plasmid is broken with a nuclease at the origin of transfer (oriT) sequence that determines where on the plasmid DNA transfer is initiated by serving as the replication start site where DNA replication enzymes will nick the DNA to initiate DNA replication and transfer. The nicked strand enters the recipient bacterium while the other plasmid strand remains in the donor. Each strand then makes a complementary copy. The recipient then becomes an F^+ male and can make a sex pilus.

In addition, mobilizable plasmids that lack the tra genes for self-transmissibility but possess the oriT sequences for initiation of DNA transfer, may also be transferred by conjugation. The tra genes of the F^+ plasmid enable a mating pair to form and the oriT sequences of the mobilizable plasmid enable the DNA to moves through the conjugative bridge.

Hfr (High Frequency Recombinant) Conjugation

Hfr conjugation begins when an F^+ plasmid with tra genes coding for mating pair formation inserts or integrates into the chromosome to form an Hfr bacterium. (A plasmid that is able to integrate into the host nucleoid is called an episome.) A nuclease then breaks one strand of the donor's DNA at the origin of transfer (oriT) location of the inserted F^+ plasmid and the nicked strand of the donor DNA begins to enter the recipient bacterium. The remaining non-nicked DNA strand remains in the donor and makes a complementary copy of itself.

The bacterial connection usually breaks before the transfer of the entire chromosome is completed so the remainder of the F^+ plasmid seldom enters the recipient. As a result, there is a transfer of some chromosomal DNA, which may be exchanged for a piece of the recipient's DNA through homologous recombination, but not the ability to form a conjugation pilus and mating pairs.

EAGLE EFFECT

The Eagle effect, Eagle phenomenon, or paradoxical zone phenomenon, named after Harry Eagle who first described it, originally referred to the paradoxically reduced antibacterial effect of penicillin at high doses, though recent usage generally refers to the relative lack of efficacy of beta lactam antibacterial drugs on infections having large numbers of bacteria. The former effect is paradoxical because the effectiveness of an antibiotic generally rises with increasing drug concentration.

Mechanism

Proposed mechanisms:

- Reduced expression of penicillin binding proteins during stationary growth phase.
- Induction of microbial resistance mechanisms (such as beta lactamases with short half-lives) by high drug concentrations.
- Precipitation of antimicrobial drug in vitro, possibly also leading to the crystallized drug being mis-detected as colonies of the microbe.
- Self-antagonising the receptor with which it binds (penicillin binding proteins, for example, in the case of a penicillin).

Penicillin is a bactericidal antibiotic that works by inhibiting cell wall synthesis but this synthesis only occurs when bacteria are actively replicating (or in the log phase of growth). In cases of extremely high bacterial burden (such as with Group A Strep), bacteria may be in the stationary phase of growth. In this instance since no bacteria are actively replicating (presumably due to nutrient restriction) penicillin has no activity. This is why adding an antibiotic like clindamycin, which acts ribosomally, kills some of the bacteria and returns them to the log phase of growth.

SELECTIVITY FACTOR

Selectivity factor is a quantifiable measure of how efficient an antibiotic is during the process of gene selection. It measures of the capacity an antibiotic to select for transfected (resistant) cells that contain a selectable marker, while killing untransfected (sensitive) cells that do not contain a selectable marker. A selectivity factor higher than 10 is optimal. This means the concentration of antibiotic is sufficient to kill untransfected cells but not toxic enough to kill transfected cells. A selectivity factor lower than 10 means the concentration of antibiotic needed for selection is too close to the toxic concentration for the transfected cells. As a result, fewer transfected cells survive and more untransfected cells survive. In this case an alternative antibiotic should be considered.

Calculating the Selectivity Factor

The method uses a modified MTT assay. The MTT assay is a colorimetric assay used to assess cell metabolic activity. The assay is based on the reduction of yellow tetrazolium salt (MTT) by active cells to produce purple formazan crystals which accumulate in living cells. Cells are lysed, the crystals are dissolved, and the absorbance of the solution is analysed on a spectrophotometer as a measure of cell viability. In situations where the use of MTT is problematic, PI or Sytox Green screening in a fluorescence plate reader can be considered. The next step is to generate a kill curve which defines the ideal concentration of a selection antibiotic to kill untransfected cells. Curves are generated for both sensitive cells and resistant cells. The half-maximal Inhibitory Concentration (IC50) can be calculated, which measures the potency of the antibiotic.

The selectivity factor is calculated as follows:

$SF = IC_{5}0R/IC_{50}S$ whereas SF = selectivity factor; IC_{50} = half-maximal inhibitory concentration; R= resistant cells; S = sensitive cells.

Advantages of the Selectivity Factor

The selectivity factor has the following advantages: It is quantitative thus can be reported numerically using a microplate reader, it streamlines the process of generating stable cell lines, it considers both sensitive and resistant cells, and it allows comparison of the consistency and quality of antibiotics from different batches, vendors, and manufacturing methods.

Practical uses of the Selectivity Factor

The selectivity factor can be used for the creation of stably transfected cell lines, an important tool in drug discovery, biomedical research, and biological pathway investigation. Cell line creation involves transfection (transferring the gene into the cell line), and selection (applying selective pressure in the form of an antibiotic. Transfection efficiency is dependent on cell type, cell density, vector, and transfection method. Selection efficiency depends on the capacity of the antibiotic to kill the parental cells but not the transfected cells.

R-FACTOR

Resistance transfer factor is an old name for a plasmid that codes for antibiotic resistance. R-factor was first demonstrated in *Shigella* in 1959 by Japanese scientists. Often, R-factors code for more than one antibiotic resistance factor: genes that encode resistance to unrelated antibiotics may be carried on a single R-factor, sometimes up to 8 different resistances. Many R-factors can pass from one bacterium to another through bacterial conjugation and are a common means by which antibiotic resistance spreads between bacterial species, genera and even families. For example, RP1, a plasmid that encodes resistance to ampicillin, tetracycline and kanamycin originated in a species of *Pseudomonas*, from the family Pseudomonadaceae, but can also be maintained in bacteria belonging to the family Enterobacteriaceae, such as *Escherichia coli*.

Transmission

Bacteria containing F-factors (said to be "F+") have the capability for horizontal gene transfer; they can construct a sex pilus, which emerges from the donor bacterium and ensnares the recipient bacterium, draws it in, and eventually triggers the formation of a mating bridge, merging the cytoplasms of two bacteria via a controlled pore. This pore allows the transfer of genetic material, such as a plasmid. Conjugation allows two bacteria, not necessarily from the same species, to transfer genetic material one way. Since many R-factors contain F-plasmids, antibiotic resistance can be easily spread among a population of bacteria. Also, R-factors can be taken up by "DNA pumps" in their membranes via transformation, or less commonly through viral mediated transduction, or via bacteriophage, although conjugation is the most common means of antibiotic resistance spread. They contain the gene called RTF (Resistance transfer factor).

References

- Pollard, Andrew J.; mccracken, George H.; Finn, Adam (2004). Hot Topics in Infection and Immunity in Children. Springer. P. 187. ISBN 9780306483448
- 3A-Horizontal-Gene-Transfer-in-Bacteria, Bacterial-Genetics, Bacterial-Genet-ics-and-the-Chemical-Control-of-Bacteria, Microbiology-(Kaiser), Microbiology: libretexts.org, Retrieved 31 March, 2020
- Delrue, I; Pan, Q; Baczmanska, AK; Callens, BW; Verdoodt, LLM (August 2018). "Determination of the Selection Capacity of Antibiotics for Gene Selection". Biotechnology Journal. 13 (8): e1700747. Doi:10.1002/biot.201700747. PMID 29436782
- Prokaryotic Cell Structure: Pili". Archived from the original on 2016-12-07. Retrieved 2017-01-19

Permissions

All chapters in this book are published with permission under the Creative Commons Attribution Share Alike License or equivalent. Every chapter published in this book has been scrutinized by our experts. Their significance has been extensively debated. The topics covered herein carry significant information for a comprehensive understanding. They may even be implemented as practical applications or may be referred to as a beginning point for further studies.

We would like to thank the editorial team for lending their expertise to make the book truly unique. They have played a crucial role in the development of this book. Without their invaluable contributions this book wouldn't have been possible. They have made vital efforts to compile up to date information on the varied aspects of this subject to make this book a valuable addition to the collection of many professionals and students.

This book was conceptualized with the vision of imparting up-to-date and integrated information in this field. To ensure the same, a matchless editorial board was set up. Every individual on the board went through rigorous rounds of assessment to prove their worth. After which they invested a large part of their time researching and compiling the most relevant data for our readers.

The editorial board has been involved in producing this book since its inception. They have spent rigorous hours researching and exploring the diverse topics which have resulted in the successful publishing of this book. They have passed on their knowledge of decades through this book. To expedite this challenging task, the publisher supported the team at every step. A small team of assistant editors was also appointed to further simplify the editing procedure and attain best results for the readers.

Apart from the editorial board, the designing team has also invested a significant amount of their time in understanding the subject and creating the most relevant covers. They scrutinized every image to scout for the most suitable representation of the subject and create an appropriate cover for the book.

The publishing team has been an ardent support to the editorial, designing and production team. Their endless efforts to recruit the best for this project, has resulted in the accomplishment of this book. They are a veteran in the field of academics and their pool of knowledge is as vast as their experience in printing. Their expertise and guidance has proved useful at every step. Their uncompromising quality standards have made this book an exceptional effort. Their encouragement from time to time has been an inspiration for everyone.

The publisher and the editorial board hope that this book will prove to be a valuable piece of knowledge for students, practitioners and scholars across the globe.

Index

www.ingramcontent.com/pod-product-compliance
Lightning Source LLC
LaVergne TN
LVHW080623210926
848485LV00003B/208